Ministry of Agriculture, Fisheries and Food

Food Advisory Committee Report on its Review of Food Labelling and Advertising 1990

FdAC/REP/10

LONDON: HMSO

© Crown copyright 1991
First published 1991

ISBN 0 11 242913 0

CONTENTS

FOOD ADVISORY COMMITTEE

During the course of this review the terms of reference of the Food Advisory Committee were:

'To advise the Minister of Agriculture, Fisheries and Food, the Secretary of State for Social Services, the Secretary of State for Wales, the Secretary of State for Scotland and the Head of the Department of Health and Social Services for Northern Ireland on matters referred to it by Ministers relating to:

(i)   the composition, labelling and advertising of food;

(ii) additives, contaminants and other substances which are, or may be, present in food or used in its preparation;

with particular reference to the exercise of powers conferred on Ministers by Sections 4, 5 and 7 of the Food Act 1984 and the corresponding provisions in enactments relating to Scotland and Northern Ireland'.

The following served on the Food Advisory Committee during this review:

Dr Ewan S Page MA, PhD, BSc, CBIM, FBCS, C Eng, FASA (Chairman)
Dr Margaret Ashwell BSc (Hons), PhD, MIEHc, FIFST, FRSH
Mr Michael Boxall ACII, CBIM (until March 1990)
Mr Dennis Cumming (from July 1990)
Professor Douglas Georgala CBE, PhD, FIFST
Dr Tom Gorsuch OBE, BSc, PhD, FIFST
Professor R Marian Hicks BSc, PhD, DSc, FRCPath
Mr Roger Manley, FITSA
Miss Patricia Mann FCAM, FIPA, CBIM, FRSA
Mrs Jill Moore OBE
Dr Bryan Nichols MSc, PhD, FIFST
Mr Ronald Nicolson M Chem A, FIFST, FIWEM, CChem, FRSC
Mrs Barbara Saunders BA (Hons), FCP
Mr Tony Skrimshire C Chem, FRSC, FIFST
Professor Paul Turner MD, BSc, Hon FIBiol FRCP, FFPM
Dr Roger C Whitehead BSc, MA, PhD, FIBiol, C Biol, Hon MRCP

Joint Secretaries

Mr J Horton (Administrative)
Dr DP Atkins BSc, PhD (Scientific)

SECTION I

INTRODUCTION

SCOPE OF THE REVIEW

1. In November 1989 we were asked by Ministers to carry out a review of labelling legislation and practices and to advise on how these might best be developed, within the context of the European Community, to give consumers what they need and want to know. The Committee was particularly asked to address the following issues:

i. whether the present and forthcoming statutory requirements represent the essential information required to inform consumers and if not what changes should be made;

ii. in addition to the advice the Committee has already offered, whether, and if so what, controls should be introduced for claims beyond those already in place;

iii. whether particular methods of rearing or slaughtering animals or poultry should be indicated on labels;

iv. whether treatments to living animals or growing or harvested crops with veterinary medicines or pesticides should be indicated on labels;

v. whether certain generic indications currently permitted (e.g. meat, oil, fat) are sufficient to meet current consumer requirements;

vi. an overall assessment of the balance between essential information on a label and other information, the sum of which might confuse consumers; whether there is a case for the standardised label;

vii. what essential information also needs to be given with foodstuffs that are sold loose; and

viii. whether any particular information needs to be controlled as to its placing, and prominence on the label/notice.

2. Because of expected developments in European Community labelling legislation, particularly in the areas of ingredient listing and claims, we were asked to complete our review as quickly as possible and to report by the end of 1990. In view of this short timescale we have focused on the particular issues identified above and these were the areas on which comments were sought from about 500 interested parties in December 1989. Almost one hundred organisations responded to the consultation letter. A list of respondents is at Appendix I. In particular we have not reviewed issues which we have recently considered, for example nutrition claims and the use of the word 'natural', where we believe our earlier advice (see appendices IV and V) holds good; neither have we addressed nutrition labelling since the UK Government is broadly content with the recently adopted European Community Directive on nutrition labelling rules. We believe that a UK review of nutrition labelling would be more appropriate once the new rules have been in operation for a few years and prior to the European Commission's review of the directive which is due five years after implementation.

CONSUMER SURVEY

3.   As well as the comments of interested parties, mainly consumer, enforcement, trade and health groups, we had the benefit of the results of a market research survey on consumers' views on labelling, including the sort of information which consumers most often look for and use, items which they find confusing and why, presentation of labelling information and additional details which they would find useful. The survey was commissioned by the Central Office of Information on behalf of the Ministry of Agriculture, Fisheries and Food. A summary of the results, including details of the sample and how the survey was carried out, is at Appendix II.

CURRENT LEGISLATIVE POSITION

4.   The main food labelling and advertising rules in England and Wales are set out in the Food Labelling Regulations 1984 (SI No 1305) with parallel legislation in Scotland and Northern Ireland. The regulations implement European Community Council Directive 79/112 on the approximation of the laws of the Member States relating to the labelling, presentation and advertising of foodstuffs. This Directive is one of the main framework directives for the establishment of a Single European Market in foodstuffs. It largely harmonises European Community food labelling legislation and further harmonisation has recently taken place with the adoption of an amending directive 89/395 covering in particular datemarking, the labelling of irradiated foods, foods delivered to caterers and particulars given in accompanying trade documents. We welcome these changes and acknowledge the considerable progress that has been made in the Community in this important area of food legislation. We fully recognise that harmonised legislation means that the UK is unable to act unilaterally to make changes to most existing food labelling legislation or to introduce new requirements at least without European Community approval and in most cases without European Community legislation. Controls on claims have been the one area which has not been harmonised and where a number of national provisions exist. However a draft Community directive is expected to be tabled shortly. Responsibility for making proposals for food labelling legislation rests with the European Commission but we would hope that the Commission and those responsible for labelling in other Member States will find our assessment helpful in determining future policy.

5.   Appendix III is a general summary of current and forthcoming food labelling legislation, including recently agreed changes and proposals under development within the UK and the European Community. Further details are contained in the individual sections of the report.

SECTION II

SUMMARY AND RECOMMENDATIONS

STATUTORY REQUIREMENTS FOR PREPACKED FOODS (SECTION IV)

6. We believe that in general the statutory requirements of the Food Labelling Regulations provide the majority of consumers with an adequate amount of information about the nature and composition of food to enable fair and informed choices of purchase to be made. However, we consider that improvements are needed in certain provisions of the Regulations. In most cases we have recommended statutory amendments or additions to the current legal requirements but where we believe regulatory action would be difficult or impractical to achieve on a European Community basis we have suggested the development of guidelines. We recognise that where Ministers have powers to regulate, guidelines produced by Government could be seen as legislating through the back door. Nevertheless we believe that if guidelines are drawn up with the full involvement of interested parties they can have a useful role to play in areas where further legislation would be difficult, or is simply not required, but where improvements in interpretation are desirable. We appreciate that there can be weaknesses in this approach, not least because guidelines are voluntary and may not be applied to imports. We suggest that the guidelines should be submitted to the European Commission so that they can consider Community-wide action.

Descriptive and customary names

7. Whilst we believe that the rules on descriptive names are satisfactory in principle, (but see paragraphs 20–21 on placing and prominence) *we recommend that non-statutory guidelines are drawn up involving trade, consumer and enforcement bodies, to ensure more consistent interpretation of the requirements.* Aspects which might particularly be addressed are: promotional elements in descriptive names, interpretative difficulties with indicating treatments or physical condition of the food, the use of similar descriptive names or trade marks for products of widely differing compositions, and criteria for determining which ingredients of a food need to be referred to in the descriptive name. *We recommend that the exemption from providing a descriptive name for a food which carries a customary name should be removed. Customary names should continue to be permitted, but not in place of a descriptive name* (paragraph 51).

Food processes and treatments

8. We consider the legislative requirements in respect of the processes or treatments a food has undergone to be satisfactory in principle. However, we believe they are not being strictly observed, leading to abuses of certain basic terms in this area. Examples of these are 'roasted', 'smoked', 'filleted'. *Advice on these should be covered in the guidelines mentioned above* (paragraph 55).

Special emphasis rules

9. We consider that the special emphasis rules fail to provide consumers with exactly the information they need about the composition of food to allow comparisons between similar foods to be made. Nevertheless *we feel that it is inappropriate to make specific recommendations for changing these rules at this stage as we believe that the forthcoming European Community proposals for quantitative ingredient declarations of the major characterising ingredients of foods, should resolve the difficulties over the present requirements. We urge progress in this area* (paragraph 59).

Place/Country of origin

10. *We do not recommend any changes in the current legal provisions.* We consider that the current rules, together with the general controls on false and misleading food descriptions, are satisfactory to ensure consumers are not misled in respect of the origin of food. Nevertheless *we believe that the guidelines on the 'name of the food', mentioned above, should include guidance on when the use of a place name in the name of the food would require the origin mark to be given* (paragraph 63).

'Flavour' and 'flavoured'

11. We consider that consumers experience difficulty in understanding the significant difference between these two words. We believe there is no ready solution to the problem but *recommend that consideration be given to prohibiting the use of the word 'flavour' and replacing it with 'taste'* which has a more distinct meaning, better understood by consumers (paragraph 66).

Minimum durability marking

12. *We consider that the current provisions of the Regulations in respect of minimum durability marking of foods are satisfactory.* We also consider that the agreed changes at European Community level in the rules for the datemarking of foods, which will introduce a compulsory 'use by' date for microbiologically highly perishable foods, remove the UK 'sell by' and introduce a reduced list of foods exempt from date-marking, are useful further developments in this area. Additional forms of date-marking foods such as 'display until', which are for the use of the retailing staff, are unlikely to confuse or mislead consumers. We therefore do not wish to recommend any restriction on their use (paragraphs 71–72).

Special storage instructions, conditions of use and instructions for use

13. *We consider that there is an urgent need for clearer and more specific storage instructions for foods and we recommend statutory changes to require the storage temperature to be given on foods for which strict storage temperatures are required to maintain safety and quality throughout the intended life of the food. Food manufacturers should be encouraged through non-statutory guidelines to provide more accurate instructions for use to ensure that microbiologically highly perishable foods remain wholesome.* The guidelines should address such aspects as the best way to wrap foods and guidance on vulnerable points in the 'shop to home' chain together with instructions, if appropriate, on freezing, thawing and cooking foods. This is particularly important if the food is likely to be microwaved (paragraphs 76–77).

Description of ingredient list

14. We looked at the suggestion that it should be made a requirement for products to state clearly on the label that the ingredients are shown in descending order of ingoing weight. However we felt that education rather than labelling was the answer. Moreover we believe that the forthcoming European Commission proposals requiring quantitative ingredient declarations should largely meet the need for further information (paragraphs 78–81).

Exemptions from ingredient listing

15. *We consider that in principle all ingredients should be listed on food labels without exception.* This is in the interests of consumer choice and in line with the general move towards more informative labelling. There is therefore a need to review the current exemptions for certain foods from ingredient listing. Forthcoming European Community proposals for ingredient listing of alcoholic drinks and single ingredient foods are to be welcomed. However, *we recommend that a full review of the exemptions provided for in European Community legislation should be sought with the aim of ensuring full ingredient listing* for these foods (paragraph 85).

25% cut-off rule for declaring ingredients of compound ingredients

16. We consider that this provision of the Regulations is unhelpful to consumers and is no longer justified because of the purely arbitrary nature of the 25% figure, and because it leaves the way open for abuse by those who may wish to avoid indicating particular ingredients on the label. *We therefore recommend that the exemption is removed from European Community legislation.* Constituents of compound ingredients should continue to be listed in the format currently required (paragraph 88).

Generic indications

17. We consider that whilst generic indications provide a useful degree of labelling flexibility to the trade, certain of these generics, in particular, 'meat', 'other meat', 'fish', 'other fish', 'oil' and 'fat' are failing to provide most consumers with sufficient information on which to make proper judgments about foods. *We consider that the current list of permitted generic indications (both national and European Community) should be reviewed with the aim of providing consumers with fuller and more accurate information by reducing the list to those generic terms which are considered to be strictly necessary to afford manufacturers the degree of flexibility they need.* We also believe that more meaningful conditions of use should be introduced to bring the terms more into line with consumers' understanding of them (paragraph 92).

Ingredient listing of additives

18. We recognise that improvements have already been achieved in the rules for the labelling of additives in food. Nevertheless *we believe that the mixing of E numbers and specific names within a single ingredients list, as permitted under the current rules, is unhelpful and confusing for consumers. We recommend that amendment to the European Community legislation is sought to prohibit this practice. We also recommend extension of the list of function names for additives together with provision for manufacturers, in the absence of a specified function, to apply a meaningful descriptive function so that the reason for all the additives used in a food is available to the purchaser* (paragraph 96).

Carry-over additives

19. We believe that the exemption from indicating a carry-over additive if it serves no significant technological function in the final food, and for processing aids, is failing to give consumers sufficient information as regards the nature and composition of food, to allow informed choices to be made. We further consider that consumers are for the most part misled under the present arrangements as they believe the requirements of the legislation to be for full, rather than partial listing of additives. *We recommend that carry-over additives and processing aids, present in the final food at a level at which we believe consumers would wish to know of their presence, should be indicated in the list of ingredients irrespective of their function.* In order to balance consumer needs against the practical difficulties for the trade *we recommend that a de minimis level should be set for declaring these substances in the final food,* and we suggest 10% of the maximum permitted level or 10 mg per kilogram whichever is the less. We consider the same principle should equally be extended to processing aids. For both processing aids and carry-over additives we recommend that, based on the typical levels of these substances in the final food, an indication such as 'includes x and y' at the end of the ingredients list should be satisfactory (paragraphs 99–100).

PRESENTATION OF LABELLING INFORMATION (SECTION V)

Standard panel

20. We consider that improvements to the present general requirements for the manner of marking foods are needed to ensure that consumers are able to obtain the essential information quickly and easily. We are impressed by the clarity of certain labels, such as those for pet foods, using a statutory box format for key information and *we take the view that a standard panel, prominently displayed on the label, would help to counter the difficulties over the relative prominence and placing of the statutory and non-statutory information. We recommend that all the statutory information, with the exception of the name and address of the responsible manufacturer, packer or seller, should be included in the box together with nutrition labelling information and any additional safety information. All other information should be excluded from the box.* To allow a degree of flexibility *we recommend that 'signposting' (for example, 'see lid', 'on back of pack') should be allowed for the following information; the date-mark, storage instructions, conditions of use, instructions for use, and nutrition labelling.* The following should always be given in the panel: the true name of the food (as opposed to the brand or fancy name), the list of ingredients; the weight; alcoholic strength if appropriate and place of origin if necessary. As regards the positioning and prominence of the statutory panel we believe that the current rules on the intelligibility of the statutory particulars fully cover this. Nevertheless we would draw industry's attention to these requirements in the Food Labelling Regulations, in particular that the information 'shall be marked in a conspicuous place in such a way as to be easily visible' and 'not in any way be hidden, obscured or interrupted by any other written or pictorial matter'. *We would encourage industry to respond to consumers' very clear concerns about the size of lettering used on food labels.* We do not wish to recommend further controls on non-statutory promotional material on labels as we believe that the statutory standard panel would provide an effective counter-balance to this material. Nevertheless we would remind industry of the general controls in the Food Safety Act 1990 on misleading labelling and presentation of food (paragraphs 107–110).

True name

21. To improve the position on the lack of prominence given to the true name of the food compared with brand and fancy names *we recommend that the true name should be required to appear in immediate proximity to the trade or fancy name most prominent under normal conditions of purchase in addition to its inclusion in the standard panel* (paragraph 111).

7

Symbols

22. *We do not at present wish to make any recommendations in respect of pictorial or symbol labelling.* This may be useful in certain circumstances but such labelling should not be a replacement for clear and precise written information. We welcome the fact that the Ministry of Agriculture, Fisheries and Food is sponsoring research into graphical representation of nutrition information and we believe that any further action should await the results of the Ministry's nutrition labelling research (paragraph 112).

CLAIMS (SECTION VI)

Use of the word 'natural' and nutrition claims

23. There are no new recommendations regarding the use of the word 'natural' or on the control of nutrition claims as we believe earlier advice (see appendices IV and V) holds good.

General principles

24. In order to provide a more consistent and readily understandable framework for legislation, *we recommend that food claims should be controlled by the following general principles* (paragraphs 121–123):

(a) a food must be able to fulfil the claim being made for it and adequate labelling information must be given to show consumers that the claim is justified;

(b) where the claim is potentially ambiguous or imprecise (for example, 'light') it must be clearly explained and justified on the label, and be capable of substantiation;

(c) a claim that a food is 'free from' a substance or treatment should not be made if all the same class or category of foods are similarly free;

(d) words or phrases which imply that a food is free from any specific characteristic ingredient or substance should not be used if the food contains other ingredients or substances with the same characteristic;

(e) meaningless descriptions should not be used;

(f) comparative claims must be justified against relative and generally applicable criteria;

(g) the label should give a sufficiently full description of the food in relation to the area for which the claim is made to ensure that selective claims, even if true, do not mislead;

(h) absolute claims must be justified against absolute criteria set for a given nutrient and applying to all foods;

(i) where a food is naturally 'low' or 'high' in a substance the claim should be 'a low/high x food'.

Medicinal claims

25. *We recommend that direct medicinal claims for food should continue to be generally banned* (paragraphs 124–125).

26. *We recommend that claims related to deficiency diseases which are not a problem in the European Community should not be permitted unless it is clear the food is intended for a particular minority in the population with special needs* (paragraph 128).

Health claims

27. We defined a health claim as 'any statement, suggestion or implication in food labelling and advertising (including brand names and pictures) that a food is in some way beneficial to health, and lying in the spectrum between, but not including, nutrient claims and medicinal claims' (paragraph 131). We have strong reservations about the use of such claims and believe that there is considerable potential for consumers to be misled or confused (paragraphs 118, 126 and 130). We believe that stringent controls are needed and *we recommend that health claims should only be permitted if they can be justified according to any recommendations that have been made or supported by the Chief Medical Officer (CMO)* (paragraph 134).

28. *We recommend that health claims should additionally be controlled by the following general principles* (paragraph 135):

(a) the claim must relate to the food as eaten rather than to the generic properties of any of the ingredients;

(b) a food, when consumed in normal dietary quantities, must be able to fulfil the claim being made for it and adequate labelling information must be given to show consumers that the claim is justified;

(c) the label should give a full description of the food to ensure that selective claims, even if true, do not mislead and any claim should trigger full nutrition labelling (at least the group of eight nutrients in the European Community Nutrition Labelling Rules Directive);

(d) the role of the specific food should be explained in relation to the overall diet and other factors.

29. We believe that the risk of consumers being misled is even greater in the case of endorsement schemes used to promote certain types of food as being beneficial to health. Such schemes require significant resources to operate in a way that is scientifically acceptable, not open to abuse, easy to monitor and not liable to cause consumer concern, and since there are no suitable existing mechanisms for introducing controls over such schemes we believe it is important to err on the side of caution. *We recommend that health endorsement schemes should not be permitted and that the Government considers other ways to achieve dietary change by improving basic nutrition education and explaining the relationship of nutrition with health* (paragraphs 136–139).

30. Endorsements by individuals (testimonials) in ways which amount to a health claim because of implied or direct references to the health-giving properties of the product can also offer considerable scope for misleading consumers. We see no problem in principle with promotion by famous personalities or sponsorship of events which may be designed to give the food a 'healthy' image. We are concerned that consumers may be seriously misled by explicit or implied claims making reference to possible disease risk factors or more generally to what the food can do to improve or maintain health. *We recommend that all such health endorsements should be banned* (paragraph 140).

METHODS OF REARING AND SLAUGHTERING (SECTION VII)

Rearing methods

31. We accept that a substantial proportion, although probably a minority, of consumers wish to have additional labelling to indicate farming/production methods for foods and believe that ethical considerations are behind the request rather than any concern for the quality of the final food produced by the various systems. We also recognise that there are considerable practical difficulties which would be associated with the general requirement for such labelling. It is arguable whether a labelling requirement is impracticable but nevertheless we take the view that concerns about animal welfare are best addressed through means other than labelling. It is our view that labelling requirements are to enable informed choices to be made about the final food itself and since there is no evidence of difference in substance or quality between foods produced from animals reared under various husbandry systems *we do not recommend a requirement to indicate animal rearing methods on food labels.*

32. We are concerned however, that consumers should not be confused or misled by terms used by some sectors of the trade who are already voluntarily providing this information in response to consumer demand and *we therefore recommend that the possibility of drawing up a standard system, which clearly defines terms to indicate welfare-orientated rearing methods, should be considered* to control this type of claim. We have noted the progress being made on European Community legislation on organic foods. *We would urge extension of the regulations to animal products as quickly as possible, and that rearing methods should be fully taken into account* (paragraphs 152–157).

Slaughtering methods

33. We recognise that, whilst few consumers wish to have information on methods of slaughter, it may be that consumers are less well informed about slaughtering methods and the associated welfare issues than they are about methods of rearing animals. In particular we believe that most consumers are unlikely to be aware that meat from religiously slaughtered animals is diverted to the general market. On the question of a specific labelling requirement we believe that the same considerations and principles apply as those we have set out in relation to labelling to indicate methods of rearing. In particular we believe that the major factor is that there is no physical difference between meat produced from religiously slaughtered animals and those slaughtered by conventional methods and that labelling is not the answer to welfare concerns, whether those concerns are justified or not. *We therefore do not recommend a requirement to indicate slaughtering methods on food labels* (paragraph 159).

TREATMENTS TO LIVING ANIMALS AND CROPS (SECTION VIII)

34. We consider that it is a minority of consumers who positively wish to have information on veterinary medicines on food labels and that the practical problems and risks of consumers being misled by such information rather than informed, suggests that *there should not be a statutory labelling requirement.* We recommend accordingly. We have also decided against a requirement for labelling of non-therapeutic treatments with veterinary medicines, as in most cases the final food would be no different from that derived from animals which had not been treated for non-therapeutic reasons, and because of the practical difficulties of meeting such a requirement. It is the Committee's view that consumer concerns over non-essential treatments such as the use of antibiotics and growth promoters should be addressed not through labelling, but by ensuring that proper controls are in place and by consumer education about those controls and about the effect (or lack of any effect) in foods and consumers of the use of treatments (paragraph 180).

35. We consider that a significant proportion of consumers would welcome some general information on the use of pesticides and post-harvest treatments (in the widest sense of these terms) although there appears to be considerably less interest in the specific treatment used. We also consider that there may be greater concern on the part of consumers in respect of post- rather than pre-harvest treatment of fruit and vegetables as consumers wish to be able to handle and prepare the treated food appropriately. In addition to the practical difficulties and costs involved with a statutory labelling requirement, which might work to the detriment of the majority of consumers, we were concerned that labelling might be potentially confusing for consumers. However, despite the difficulties, *we feel there is merit in distinguishing between treatments such as waxes on fruit which are purposefully added post-harvest to protect the properties of the final food as sold to the consumer and those pesticides which occur adventitiously from pre-harvest application. We believe that the former should be indicated on labels or, in the case of non-prepacked foods, notices as appropriate.* Declarations should be along the lines of 'post-harvest pesticide/treatment used.' We believe that the positive and controlled labelling of organic produce will also help to ensure consumer choice in this area. *We further recommend that information to allow consumers to prepare appropriately produce which has been treated post-harvest, for example by peeling or washing thoroughly, should be given to consumers as should any special conditions of use or instructions for use* (paragraphs 184–186).

36. *We believe that public concern about pesticide treatments should be addressed through a high profile widespread education campaign* with action in schools and at European Community level. As with veterinary medicines, consumers should be encouraged to understand that unless otherwise stated the food would have been produced using pesticides in a conventional agricultural system and that such pesticides should have been used safely. Consumers should also be educated to help them understand the meaning of declarations of post-harvest treatment (paragraph 187).

FOODS PRODUCED USING GENETIC MODIFICATION (SECTION IX)

37. The labelling of foods produced using genetic modification was not specifically considered as part of this review since consideration of this very important issue was already underway. The guidelines which we have drawn up to help us consider the need for special labelling on a case by case basis are included at Appendix VI.

NON-PREPACKED FOODS INCLUDING FOODS SOLD BY CATERERS (SECTION X)

Non-prepacked retail sales

38. There is no reason in principle why the information required to be given for prepacked foods should not be extended to cover all foods. We believe that, in principle, the present labelling requirements are unsatisfactory as they do not represent the essential information which most consumers wish to receive about food sold this way. However we accept that there would be practical difficulties for some sectors and believe that consumer needs could be met by an intermediate level of labelling to give certain key particulars such as date-mark, storage instructions, conditions or instructions for use according to the particular type or category of food on offer. *Whilst we decided against recommending further statutory controls* in this area, as these would not be flexible enough to allow the trade to choose the most essential and appropriate labelling information for the particular foods on offer, and the means of conveying that information to the consumer, *we nevertheless strongly urge the trade to respond positively and fully to consumers' wishes for more labelling information for non-prepacked foods* (paragraphs 199–200).

Additives

39. We consider that the present requirement for certain categories of additives to be indicated for non-prepacked foods can no longer be justified as it is failing to provide consumers with the full and clear information about additives in foods which they wish to have. It is likely that many consumers believe the list of additive categories indicated for these foods to be a comprehensive one and are therefore misled by partial information. *We therefore recommend that the present requirements of the Regulations for certain categories of additives to be indicated should be extended to require all categories to be listed.* Where there is no appropriate category name specified in the Regulations *we recommend,* in line with the recommendations we have made in respect of labelling of additives in food, *that a meaningful descriptive function name should be given. We further recommend that where the retailer, in response to our recommendation for fuller labelling of non-prepacked foods, voluntarily provides more detailed information including a list of ingredients, then any additives present should be indicated in line with the requirements of the Regulations in respect of listing additives in prepacked foods* (paragraph 201).

Foods prepacked for direct sale

40. We believe that the present requirements for the labelling of foods prepacked for direct sale are out of step with developments in food sold this way so that an anomalous situation has arisen as regards the amount of labelling information provided to consumers purchasing very similar foods, often in close proximity to each other in the shop. *We recommend that the requirements of the labelling of foods prepacked for direct sale should be extended to the same level as for prepacked foods.* In recognition of the practical difficulties, particularly as regards providing a detailed list of ingredients in order of ingoing weight, we recommend that the extended labelling requirements should be framed so as to allow 'typical value' indications of the ingredients for such products (paragraph 202).

Food sold in catering establishments

41. For prepacked foods sold by caterers, such as sandwiches, filled rolls and other similar bread products, *where the Regulations already require full labelling information when these products are supplied to the caterer, we consider that this information should be passed on to the consumer.* We do not however recommend further statutory measures. For other catering foods we do not consider that changes to the legislation are needed although *we strongly urge caterers to provide the fullest possible descriptions of the food on offer to the consumer* (paragraph 203).

SECTION III

PHILOSOPHY OF LABELLING

BACKGROUND

42. We have given some thought to the aims and objectives of food labelling as necessary background to our review and to provide a coherent strand to our conclusions and recommendations. Whereas advertising and publicity have their place in informing consumers, it is the label or the notice accompanying the food that probably has the most immediate effect since it is here that important information needs to be communicated in order to assist comparison, selection and purchase at the point of sale and the safe storage and use of the product thereafter. We believe that current food labelling legislation, together with the changes in the pipeline (for example on date-marking, nutrition labelling and quantitative declarations of ingredients) provide considerable information and protection for consumers, but that there is room for improvement. Certain gaps have been identified (for example in so far as the legislation applies to foods sold loose) and these need to be considered. In addition there can be differences of interpretation and whilst the letter of the law may be adhered to, sometimes what we perceive to be the spirit of the law is not. Enforcement can be a problem in this subjective area if legislation is not clear and also because of the requirement of the criminal law that offences have to be proven beyond all reasonable doubt.

CONSTRAINTS

43. There are two important constraints on information provided on food labels. First, a label has a limited size and, if information is to be easily understood and clearly legible as the regulations require, only so much information can be included. Second, consumers can usefully absorb only a certain amount of information from a label or notice at the point of sale and to go over that limit not only has little value in reaching a purchasing decision but can divert attention from the key items of information. Since there are many ideas on what information might be included on a label there has to be a rigorous appraisal of all of the information given: that which is required by statute, that which is desirable for consumer information and that which we acknowledge is necessary to meet the legitimate demands of the marketing of the product.

AIMS

44. We consider that the objective of food labelling is to provide sufficient information about all food available for purchase (not just pre-packed foods) in a clear manner so that the consumer can make informed choices and can store and prepare the food appropriately. Labelling controls must:

- specify what information is essential;
- ensure that information is readily comprehensible;
- address special needs on products provided for particular sectors of the population;
- prohibit information or material which is likely to mislead or confuse consumers;
- cater for significant minority interests, so long as the majority of consumers are not likely to be confused or overloaded by the additional information.

45. We firmly believe that labelling should not be used as a substitute for education. Labelling information may provide important back-up to educational campaigns (for example, nutrition labelling in relation to promoting a healthy diet) but it should not be regarded as a primary educational tool because of the constraints mentioned above and the risks of conveying over-simplistic, misleading messages in such a limited format.

SECTION IV

MAIN CURRENT STATUTORY REQUIREMENTS FOR PREPACKED FOODS

BACKGROUND

46. We considered whether the main, current statutory requirements for the labelling of prepacked foods, particularly the requirements of the Food Labelling Regulations 1984, represented the essential information required to inform the consumer. The representations we received from interested parties and the consumer survey lead us to believe that, generally speaking, the main requirements provide the majority of consumers with an adequate amount of information about the nature and composition of food to enable fair and informed choices of purchase to be made. However, respondents did identify certain areas where we consider improvements are needed. These are discussed below. In most cases we have recommended statutory amendments or additions to the current legal requirements but where we believe regulatory action would be difficult or impractical to achieve on a European Community basis we have suggested the development of guidelines. We recognise that where Ministers have powers to regulate, guidelines produced by Government could be seen as legislating through the back door. Nevertheless we believe that if guidelines are drawn up with the full involvement of interested parties they can have a useful role to play in areas where further legislation would be difficult or is simply not required but where improvements in interpretation are desirable. We appreciate that there can be weaknesses in this approach, not least because guidelines are voluntary and may not be applied to imports. We suggest that the guidelines should be submitted to the European Commission so that they can consider Community-wide action.

47. In our consideration of those requirements concerned with exemptions from standard ingredients listing provisions (paragraphs 82–100), we recognised the change that has taken place in consumer attitudes. In the past consumers' desire for information related largely to concern about possible adulteration and substitution whereas now consumers are mainly interested in the details of composition because they want to know what is in the food. We acknowledge consumers' right to know and believe that in principle all ingredients of all foods should be declared. Partial listing risks misleading or confusing consumers and this has also been taken into account in our assessment. However for practical reasons and because we believe consumer interest is focussed on ingredients purposefully added to the food and thus directly relevant to its nature, substance and quality, we distinguished between ingredients added to the food for a purpose and those components present adventitiously where it seems appropriate to apply a minimum cut-off point below which declarations would not be required. This approach also applies to our consideration of treatments applied to crops, in Section VIII.

DESCRIPTIVE AND CUSTOMARY NAMES

Legislative position

48. The Regulations require foods to give either a name prescribed by law (wholemeal bread, marmalade, salmon), a customary name, that is one which is familiar to the public but which is not necessarily directly descriptive of the food (pizza, Bakewell tart) or a descriptive name, that is one which is precise enough to indicate the true nature of the food and allow it to be distinguished from similar foods (spaghetti hoops in tomato sauce). If necessary the name should include a

14

description of its use. The Regulations also specify that the name must include an indication of the food's physical condition (powdered, concentrated, etc) and indicate any treatments the food itself has undergone (dried, frozen, smoked etc.), where omission of such information could mislead the consumer. The brand or fancy name may not substitute for the descriptive name. In addition the general provisions of the Food Safety Act 1990, and Trade Descriptions Act 1968 provide controls over misleading or false descriptions of food.

Views of interested parties

49. The majority of respondents indicated that the legislation was satisfactory in principle, but that variations in interpretation had led to problems, to the extent that products with quite different compositions could be marketed under similar descriptive names. In these circumstances it was felt that too much reliance was placed on the ingredients list to qualify inadequate descriptive names. A particular criticism was that the emphasis was too frequently placed on promoting the product, or highlighting particular 'attractive' qualities of the food rather than giving a factual description of the product. It was felt that increasingly, the distinction between the descriptive name and the brand or fancy name was not sufficiently clear. Particular concern was expressed about the 'customary names' provisions of the regulations given the move towards more informative labelling for consumers and away from compositional standards for foods. A clear majority favoured discontinuing them and wished to see a requirement for all foods to carry a descriptive name unless there was a name prescribed by law.

Consumer survey

50. The consumer survey indicated that the problems with descriptive and customary names might not be as significant as representations would suggest. Very few consumers (4%) said that they found descriptive names for products particularly confusing or unclear.

Options considered and recommendations

51. We consider that the designation used for a product, whether this is a name prescribed by law, a descriptive name or a customary name, is particularly important in initially indicating to the consumer the true nature of the food. We believe that the requirements of the legislation in this respect are broadly satisfactory although *we recommend that further clarification of the provisions through non-statutory guidelines, involving trade, consumers' and enforcement bodies, would aid more consistent interpretation.* The guidelines could address, for example: promotional elements in descriptive names, interpretative difficulties over indications of treatment or physical condition, the practice of using the same or similar descriptive names or trade marks for widely differing products, criteria for determining which ingredients of a food should be referred to in the descriptive name. As regards the exemption from giving a descriptive name for foods which carry a customary name, we consider that this exemption is no longer justified. *We recommend therefore that the exemption should be removed. Customary names should continue to be permitted but not in place of the descriptive name.*

FOOD PROCESSES AND TREATMENTS

Legislative position

52. The Regulations require the name of the food to include, or be accompanied by, an indication of its physical condition or treatment where a purchaser could be misled by the omission of that information. Special rules apply to frozen meat and offal, tenderised meat, processed peas and food frozen with dichlorodifluoromethane. Any descriptions, including a reference to a treatment or process, would also need to be an accurate statement about the condition of the food. The emphasis is on the treatment or process the food itself has undergone as it is generally not a principle of the legislation that treatments or processes applied prior to the food stage, should be indicated on the label. Voluntary indications of such information may be given.

Views of interested parties

53. Enforcement and consumer interests indicated that these provisions were satisfactory in principle, but that problems arose in practice because of less than strict compliance by the trade, together with subjective interpretational difficulties over physical condition or treatment. They suggested that the trade might be encouraged to apply the provisions more strictly and consistently. European Community standardisation of the terms used to describe the processes and treatments was proposed by some consumer interests.

Consumer survey

54. This was not specifically addressed in the consumer survey. However 45% of survey respondents thought that they would refer to and use greater detail on the general descriptions of food. The need for consumers to know exactly what they are buying, so that they can make informed choices of purchase, was the main reason given for seeking this information.

Options considered and recommendations

55. We recognise that because there is an immense and increasing number of food processes and treatments, and because individual foods may go through a number of processes at different stages of manufacture, it would be very difficult to move away from the present non-specific requirement of the regulations. For the same reasons we consider that agreement at European Community level of definitions of each food process or treatment is likely to prove impracticable. Nevertheless, whilst acknowledging these difficulties, we consider that there is room for improvement in this area as there is no doubt that certain familiar processes, when referred to in the labelling of food, are often far removed from their conventional meanings and that consumers are consequently misled. *Whilst we conclude that there is no need for legislative changes, we would draw industry's attention to these legal requirements and highlight what the Committee believes to be abuses of certain basic terms in this area.* Examples of these are:

> 'roasted' – the process as currently applied to some meat joints can mean steam cooking, followed by flash roasting and sometimes the addition of the traditional roasted colour by painting or spraying. An accurate description would be: 'steam cooked and flash roasted'.

> 'smoked' – the process which is applied in some cases is in fact a flavouring process consisting of painting, spraying, or immersing the food with or in a solution which imparts the flavour. An accurate description would be: 'smoked flavour' or 'smoky taste'.

> 'filleted' – sometimes used to describe de-boned pieces of fish which are frozen into blocks (often with the addition of polyphosphate solution) and then sawn into slices of fish. An accurate description would be: 'de-boned, frozen, reformed, sliced fish'.

As we have mentioned in paragraph 51 above, the guidelines we have recommended on the 'name of the food' provisions could include guidance on processing terms.

SPECIAL EMPHASIS RULES

Legislative position

56. The Regulations require that, where the presence or low content of an ingredient is given special emphasis, an indication of the minimum or maximum percentage, as appropriate, of that ingredient must be given either next to the name of the food or in the list of ingredients in close proximity to the name of the ingredient in question. A reference in the name of a food to a particular ingredient does not itself constitute special emphasis, for example, 'lemon sponge cake' or 'strawberry dessert'. A description, such as 'chicken and leek pie with extra chicken' would however need to take account of the special emphasis rules. It is expected that European Community proposals for quantitative ingredient declaration for the main characterising ingredients of foods will replace the current special emphasis provisions.

Views of interested parties

57. Enforcement and consumer interests were concerned that the special emphasis rules were not sufficiently clear, and that the trade did not always take account of these. The main criticism was of the provision allowing a reference to be made to a particular ingredient in the name of the food without triggering the special emphasis rules. The suggestion was that consumer needs would be better served if this particular provision was removed or reworded such that, when an ingredient or constituent was emphasised on the label, and/or appears in the name of the food, the maximum or minimum level should be declared. Then the consumer would be immediately informed that, for example a 'cod fish pie' contained, as its fish component, only 70% of cod as opposed to 100% of cod, which might be the logical assumption by the consumer. Respondents acknowledged that the inadequacies in the present special emphasis rules needed to be considered against the expected proposals for quantitative ingredient declarations which it was felt, might deal with the problems that respondents had identified.

Consumer survey

58. Given the choice of various types of additional information which would be given on food labels, over half (52%) of the respondents stated that they would refer to, and use, information on the quantities of the main characterising ingredients in foods. This placed it at the top of the list of types of additional information that consumers would wish to have. The need to know exactly what food contains was the main reason given for selecting this item. It was also interesting to note that 80% of all respondents agreed with the statement that 'I would like the quantities of main ingredients to be given', with only 8% disagreeing.

Options considered and recommendations

59. We recognise that the majority of consumers believe that they would benefit from having greater detail on labels about the quantities of the main ingredients in food, so that they are more accurately informed about the composition thus allowing better comparison between similar foods to be made. We also believe that the current special emphasis rules are failing to provide consumers with the information they need in this respect. *However, we decided that it would be inappropriate to*

17

make specific recommendations to change the current special emphasis rules at this time as we believe that the forthcoming European Community proposals for quantitative declarations of the main characterising ingredients of foods (QUID) should resolve the concerns of interested parties. We welcome the European Commission's commitment to make proposals for quantitative declarations and urge progress in this area.

PLACE/COUNTRY OF ORIGIN

Legislative position

60. The Regulations require foods to give the place of origin if omission might materially mislead a purchaser as to the true origin of the food. In these circumstances Devon Fudge, made in Germany or Newcastle, or Blackpool rock made in Cornwall, would most probably need to indicate the true origin on the label. Bakewell tarts made in Scotland or canned imported fruits with no particular origin suggested on the label would probably not need to indicate the place of origin. In practice many food manufacturers, as a matter of marketing policy, voluntarily give information about the origin of the food. Similarly information about origin may be given for promotional purposes, for example 'Welsh lamb', 'smoked Scottish salmon'. There is nothing to prevent such indications provided the information is true and does not mislead the consumer. The provisions of the Regulations are deliberately framed in these terms as the origin of a product is generally not an essential consideration in respect of the nature, substance or quality of a food, and would only become relevant if there is a possibility of the purchaser being misled about the food's true origin.

Views of interested parties

61. Some interest was expressed among respondents in extending the origin marking requirements to require the country/place of origin to be indicated at all times on the label, although it was recognised that this would have more significance for some foods than others. Trade interests suggested that this change in the labelling requirements might help those Member States seeking legislation on food quality linked to origin. Consumers saw it as a means of widening consumer choice.

Consumer survey

62. The consumer survey indicated that place of origin was the least important item of information for consumers with only 2% of respondents stating that they would specifically look for this on the label. When shown a prompt list of the types of information that appear on food and drink labels, more respondents (18%) selected this information although they still placed it well down the list of items they would particularly look for on the label.

Options considered and recommendations

63. We consider that the consultation exercise and the consumer survey both show that information about place of origin is among the least important to enable consumers to make informed choices. We believe that the current rules on origin marking, together with the general controls on false and misleading food descriptions, are satisfactory to ensure that consumers are not misled in respect of the origin of food. *We therefore recommend that no change is required in the current* *legal provisions but that the guidelines on the 'name of the food', recommended in paragraph 51 above, should include guidance on when consumers might be misled by the inclusion of a place name in the name of the food thus triggering the requirement to declare the origin of the food.*

'FLAVOUR' AND 'FLAVOURED'

Legislative position

64. The Regulations set down conditions for use of certain words and descriptions, to ensure that they are not used in a way which could mislead. Descriptions of food, implying or stating that they have the flavour of the food named in the description are controlled in this way. If a food is described as 'strawberry flavoured' the flavour must come wholly or mainly from the named food. The word 'flavour' preceded by the name of the food, for example 'strawberry flavour', may be used when the flavour is not derived from the named food. A pictorial representation of a food, implying that it has the flavour of that food, cannot be used unless the flavour comes wholly or mainly from the food in the picture.

Views of interested parties

65. Concern was expressed that these descriptions were not well understood by consumers and that they were inadequate to inform the consumer about the true nature of the food. For example 'strawberry mousse' could be used to describe products flavoured in various ways – natural flavours only, fruit and artificial flavours, fruit and natural flavours. Respondents suggested that these provisions should be reviewed and the description 'flavoured' should have conditions of use applied.

Options considered and recommendations

66. The Committee agrees that there is a problem in understanding the significant differences in labelling terms, between the meaning of these two similar words. Whilst the Committee does not believe there is a ready solution to this problem it believes that the situation would be improved if two more distinct words were used. *We consider the use of the word 'flavour' might be prohibited and replaced by 'taste'.* 'Taste' does of course have a different meaning but the concept should be well understood by consumers. *We therefore recommend that this approach sh[?] explored with the intention of amending the regulations along these lines.*

MINIMUM DURABILITY MARKING (DATEMARK)

Legislative position

67. The Regulations require most foods to carry an indication of m[?] durability. This is the period for which a food can reasonably be expected [?] its specific properties, if properly stored. The usual form of date-markin[?] before' together with an indication of the period after purchase during which the product would remain at its best. As an alternative, food with a shelf life of six weeks or less may carry a 'sell by' date followed by the day and the month, which is the latest recommended date of sale of the food. Whatever form of date-mark is used, any storage instructions which need to be observed must also be given. Certain foodstuffs are exempted from date-marking, these include long life foods, fresh fruit and vegetables, deep frozen foods and certain alcoholic drinks.

68. Under European Community Directive 89/395/EEC adopted in June 1989, important changes to the current date-marking procedures have been agreed and will be introduced on 1 January 1991. These include a compulsory 'use by' date for microbiologically highly perishable foods which after a short period could constitute an immediate danger to human health. 'Best before' would continue to be the required format for all other foods. To ensure consistency of approach, guidelines have been issued which set out the criteria to be used in determining whether the 'use by' date applies. There will be new offences for selling out-of-date 'use by' foods and for anyone other than the person responsible for the food to change the date-mark. A reduced and uniform European Community list of exempted foods is to be introduced. Long life, frozen foods, ice cream (apart from individual portions) and partially ripened cheeses (all currently exempt in the UK) will be removed from the list with effect from 20 June 1992.

Views of interested parties

69. Respondents were critical of the present exemptions from date-marking for certain foods, particularly long life and frozen foods. Concern was also expressed that the present system, which permits an optional use of 'sell by' for perishable foods, served more to confuse rather than inform consumers. The selling of out-of-date foods was also criticised. Respondents have acknowledged however that these deficiencies in the legislation would largely be resolved under the forthcoming changes in date-marking procedures.

Consumer survey

70. The consumer survey found that a date-mark on foods was the most important and most used information on the label, with 79% of respondents indicating that they particularly looked for this. On each occasion respondents placed the date-mark significantly above what was considered to be the next most important piece of labelling information for the consumer, namely the ingredients list. Very few consumers (2%) said they found the date-mark particularly confusing, although a third agreed with the statement that "I'm not sure of the difference between 'best before' and 'use by' dates". Coded and obscured date-marks were mentioned as giving rise to most confusion.

Options considered and recommendations

71. We recognise that an indication of minimum durability plays a very important part in ensuring that the consumer can eat, or use, food at the optimum time for condition and freshness. We believe that the consultation exercise and consumer survey have clearly shown that consumers make considerable use of this information and are not significantly misled or confused under the current provisions of the regulations in this respect. *We also consider that the agreed changes at European Community level, which will introduce a compulsory 'use by' date for microbiologically highly perishable foods, remove the UK 'sell by' date and introduce a reduced list of exempted foods, together with the new national offences, bring welcome improvements. We therefore conclude that no further action is necessary.*

72. The Committee also looked at the additional forms of date-marking such as 'display until' which are intended as an instruction to retailing staff rather than the consumer, and whether such systems might mislead or confuse consumers. It is our view that consumers are generally familiar with the additional systems and that there should be no problem.

SPECIAL STORAGE INSTRUCTIONS, CONDITIONS OF USE AND INSTRUCTIONS FOR USE

Legislative position

73. If it would be difficult to use a food properly without instructions, the Regulations require these to be clearly given on the label. For instance a cake mix or concentrated drink would normally need preparation instructions. Similarly instructions would be required for foods that are to be cooked in a microwave oven. Any special conditions of use must also be indicated. For example some fat spreads are unsuitable for frying foods, whilst other foods may not be suitable for freezing. The label must also indicate any special storage conditions which need to be observed if the food is to remain in good condition. Very often the storage conditions will be linked to the date-mark on the food, as the date will be set by food manufacturers on the assumption that the food will be properly stored. Examples are 'store in a cold place', 'refrigerate after opening'.

Views of interested parties

74. Respondents indicated that the existing requirements were generally satisfactory. A clear concern among respondents however was the need for more specific temperature storage information on foods, for example 'store in a refrigerator at 5°C' rather than 'keep refrigerated', or 'keep cool – store below X°C' rather than 'store in a cool place'.

Consumer survey

75. Of the 1,028 people interviewed in this survey only six identified storage instructions, and seven cooking instructions as confusing. An assortment of reasons including lack of specific temperatures, confusion on microwave/conventional cooking instructions, and inaccurate storage or cooking information was mentioned as being at the root of the confusion.

Options considered and recommendations

76. We recognise and have sympathy with the concerns expressed by many respondents to the consultation exercise that there is a need for clearer and more specific storage instructions on foods. This is a very important area of food labelling, helping to ensure that the consumer plays his or her part in maintaining food safety and quality. We believe that instructions such as 'keep cool' or 'store in refrigerator' are not helpful to many consumers and are open to misinterpretation. *We therefore recommend that, if strict storage temperatures are required to maintain food safety and quality throughout the intended life of the food, the precise temperature details (e.g. 'keep refrigerated below 5°C') should be required to be given on the label.* The necessary statutory changes should be introduced urgently. However we recognise that action at European Community level is required. We would therefore recommend in the meantime that industry should be encouraged to give storage temperatures voluntarily now.

77. *We believe that improvements are also necessary in the instructions for use currently given on labels so that consumers have more specific and more accurate information to ensure that microbiologically highly perishable foods remain good and wholesome until the time they are eaten.* Industry might find it helpful to have guidance on how best to achieve this. Aspects such as the best way to wrap foods and guidance on the vulnerable parts of the 'shop to home' chain could all be covered, together with instructions, if appropriate, on freezing, thawing and cooking. This is particularly important if the food is likely to be microwaved. We welcome the developments that are taking place through the Ministry of Agriculture, Fisheries and Food Microwave Working Party which is devising a simple labelling system that will apply both to foods intended for microwaving and to the ovens to ensure that they are used safely. We hope that every effort will be made to ensure swift progress.

DESCRIPTION OF INGREDIENT LIST

Legislative position

78. The Food Labelling Regulations require the list of ingredients, which is in descending order by ingoing weight, to be headed or preceded by an appropriate heading which consists of, or includes, the word 'ingredients'.

Views of interested parties

79. Some consumer groups suggested that it should be made a requirement for products to state clearly that the ingredients are shown in descending order of ingoing weight since consumers may not always appreciate that this is already a requirement of the Food Labelling Regulations, and may therefore not be able to take full advantage of the labelling information, particularly to distinguish between similar foods.

Conclusion

80. There is nothing to prevent the trade taking up this suggestion. It is usual practice for the trade to use the single heading 'ingredients' or the phrase 'ingredients list'. It would not be possible to introduce a compulsory requirement for the expanded heading to be used without approaching the European Commission to seek their co-operation to taking action at Community level, as the requirement would be more restrictive than the terms of the Food Labelling Directive.

81. We considered whether the provision of a heading along the lines of 'ingredients in descending order of ingoing weight' or 'ingredients – greatest first' would remove any confusion. However we felt that education was the way forward. Proposals to amend the EC Labelling Directive in respect of the quantitative ingredient declaration (QUID) are expected shortly and we felt that these should largely meet the need for further information in this area.

EXEMPTIONS FROM INGREDIENT LISTING

Legislative position

82. The Regulations require most prepacked foods to carry a complete list of ingredients, generally in descending order by in-going weight. The name used for an ingredient must generally be the name which could be used if the ingredient were being sold as the food itself. Ingredients which need not be named include: constituents of an ingredient temporarily separated in the making of the food and reintroduced in the original proportion; additives contained in an ingredient and serving no significant technological function in the finished product; additives used solely as processing aids; and any substance other than water used in an essential quantity as a solvent or carrier for an additive. There are a limited number of foods which need not bear a list of ingredients. These include all unprocessed fresh fruit and vegetables, cheese, butter, fermented milk and cream, alcoholic drinks and single ingredient foods. In addition certain products such as honey and chocolate, which are subject to their own legislation, are exempt from ingredient listing. The European Commission is committed to introducing, before 1992, proposals for ingredient listing of alcoholic drinks and single ingredient foods.

Views of interested parties

83. Consumer and enforcement interests generally took the view that in the interests of consumer choice the exemptions from the general requirement to list ingredients should be kept to the absolute minimum. Exemptions for certain foods such as chocolate and honey, permitted under specific legislation, were particularly criticised as were the general exemptions in the regulations in respect of single ingredient foods and alcoholic drinks.

Consumer survey

84. More than half (52%) of respondents identified the list of ingredients as information they particularly looked for on the label, with only the datemark coming higher on the list of priorities. The list of ingredients was however, most frequently mentioned by respondents (11% overall) as particularly confusing. Problems with understanding E numbers for additives and chemical names, followed by lack of full information on ingredients, or no listing of ingredients were the most commonly mentioned reasons for this.

Options considered and recommendations

85. We have sympathy with the view expressed by many respondents that there is a need to review the current exemptions for certain foods from the general requirement of the legislation to indicate ingredients. This is clearly in line with the move towards more informative labelling for consumers and away from strict compositional standards. We fully share the view that the exemptions should be kept to the minimum and that these should be permitted only where there are exceptional reasons for making the exemptions. *We consider that the forthcoming European Community proposals for ingredient listing of alcoholic drinks and single ingredient foods will go some way towards meeting the concerns of those who commented on the current exemptions and we welcome these proposals.* We also acknowledge that many of the exemptions from ingredient listing contained in the regulations derive from European Community Directives. *We therefore recommend that a full review of the exemptions provided for in European Community legislation should be sought with the aim of ensuring full ingredient listing for those foods.*

25% CUT-OFF RULE FOR DECLARING INGREDIENTS OF COMPOUND INGREDIENTS

Legislative position

86. The Regulations permit compound ingredients, which make up less than 25% of the final product, to be indicated by reference to the compound ingredient itself, rather than by listing the individual constituents of the compound ingredient. For example smoked sausage in a cheese and sausage pizza need only be indicated as 'smoked sausage' in the ingredients list if it constitutes less than 25% of the pizza. However, any additive present in the compound ingredient must be indicated on the label.

Views of interested parties

87. Consumer and enforcement interests were of the view that this rule had outlived its usefulness and needed to be reviewed or preferably removed. It was suggested that there was no logical reason for retaining this exemption which was based on an arbitrary cut-off figure. Consumers particularly expressed concern about the difficulty the rule presented for those intolerant to certain food ingredients and who might need to avoid the food. There was recognition among respondents that the 25% rule also led to the anomalous situation where an ingredient, added individually to the food but in very small quantities, would need to be declared on the label, whereas the same ingredient present in larger quantities but added as part of a compound ingredient, would not need to be declared. There was some concern that the lack of a definition of a 'compound ingredient' was open to abuse.

Options considered and recommendations

88. *We can find little justification for retaining this exemption which is based on a purely arbitrary figure of 25%, and which leads to the anomalous situation with regard to the level at which a particular ingredient may need to be declared on the label.* We also share respondents' concern that the provision, without a clear and detailed definition of a compound ingredient leaves the way open for abuse by those who may wish to avoid indicating particular ingredients on the label. We acknowledge that the removal of the exemption may, in certain cases, lead to longer and more complex ingredients lists for foods, but we consider that this drawback must be carefully weighed against the obvious benefits to the consumer of receiving more consistent and fuller information about the composition of food. *We therefore recommend that the exemption is removed from the legislation.* Constituents of compound ingredients should continue to be listed horizontally, as currently required, next to the compound ingredient.

GENERIC INDICATIONS

Legislative position

89. The Regulations permit certain generic names to be used in the list of ingredients, subject to certain conditions. Some of these generic names such as 'fat', 'fish', 'cheese', 'poultry meat', 'starch', 'spices and herbs', are contained in the European Community Food Labelling Directive, but others such as 'vine fruits', 'nuts', and 'meat' are permitted national generics. These indications allow a degree of flexibility to the trade to take account of changes in recipes arising from seasonal variations in supply or cost in certain raw materials, without the need to change the label on each occasion. The European Commission is committed to reviewing the list of generics with a view to producing a standard, and probably restricted, Community list.

Views of interested parties

90. Trade interests generally believed that these provisions of the regulations were adequate and felt that the flexibility they provided in allowing recipes to be changed at short notice was extremely important to prevent increases in manufacturing costs and thus in the price to the consumer. The generics 'fish', 'other fish', 'meat', 'other meat', 'oil' and 'fat' were particularly identified as important to the trade. There was some acknowledgement that the terms 'meat' and 'fish' in particular, no longer provided sufficient information for many consumers who wished to know the species of meat, whether it was muscle meat as opposed to offal, fat or gristle, and whether it had been processed in any way. It was also recognised that the origin and condition of oils and fats was important to many consumers. A specific suggestion was that fish oils should be separately identified.

91. Enforcement and consumer interests overwhelmingly took the view that generic indications, in particular the descriptions 'oil', 'fats', 'meat' and 'fish' did not give consumers sufficient information to enable them to make a proper judgement. People with food allergies or those who wish to avoid certain foods for religious or ethical reasons were mentioned as being particularly disadvantaged under the present arrangements. However, it was also felt that the majority of consumers now simply wanted fuller labelling and found these particular generics unacceptable, not least because certain of the associated definitions are too wide and not in line with consumer understanding of these terms. There was also concern that generic labelling might be used to disguise the presence of low grade ingredients and certain processed ingredients such as mechanically recovered meat. The argument about flexibility in the manufacturing process and labelling was countered by enforcement interests who suggested that this was not borne out by current practices where manufacturers often opted for specific labelling when it was to their advantage.

Options considered and recommendations

92. We agree with the majority view among interested parties that the principle of generic indications for ingredients is useful in certain limited circumstances but that certain of these generics, in particular 'meat', 'other meat', 'fish', 'other fish', 'oil', and 'fat' are failing to provide most consumers with sufficient information on which to base their choice. *We therefore welcome the European Commission's commitment to review the current list of permitted generic indications and recommend that the list should be reduced to what is considered to be really necessary to afford manufacturers the degree of flexibility they need, and where possible providing definitions of these terms which are more meaningful to consumers.*

INGREDIENT LISTING OF ADDITIVES

Legislative position

93. The Regulations require additives used as ingredients (except flavourings) to be described in the ingredients list using the appropriate category name of the function which the additive performs in the food, for example 'colour', 'stabilisers', 'preservatives', followed by the specific name or serial name or both. Where no category name is available for the function performed by an additive, the additive must be declared in the ingredients list by its specific name.

Views of interested parties

94. All interests believed that the improvements introduced under the Food Labelling Regulations 1984 in the procedures for the listing of additives were to be welcomed, but that practical experience had shown that some further improvements were needed. Two problem areas were identified. First, it was felt that the mixing of E numbers and specific names in a single ingredient list was generally unhelpful to consumers and could confuse rather than inform. It was suggested that no one system was necessarily better than the other but that consistency within a single ingredients list should be encouraged. It was recognised that a requirement to indicate both E numbers and specific names would be overburdensome on the trade. The second concern was with additives not performing one of the specified functions

25

which are permitted to be indicated by the specific name alone. It was felt that consumers should have the fullest information possible about additives in food, and that consumers found it unhelpful and confusing when numbers or technical specific names were given without an indication of why these additives had been used. All interests agreed that a clear, meaningful function name should always be given.

Consumer survey

95. The list of ingredients was most frequently mentioned by consumers as particularly confusing. The most commonly mentioned reason for this was a problem with E numbers. Over half of those who found the ingredients list difficult to understand stated that they found the numbers confusing or could not understand them. A considerable number were also confused by the chemical names used and complained that the listing was not full or specific enough.

Options considered and recommendations

96. We recognise that, despite the definite improvements in the rules for the listing of additives, consumers continue to be confused over certain chemical names and E numbers, and in particular find the mixing of the two systems in a single ingredients list a significant problem. We believe that, as one system appears to have no definite advantage for consumers over the other, it would not be sensible or reasonable to restrict manufacturers' choice in this respect. *We consider however that the two systems should not be mixed within a single ingredients list.* We recognise that there may be a disadvantage to consumers in that similar foods might not then have easily comparable ingredients lists, but *the Committee does not believe that this is a significant problem. We therefore recommend that amendment to the Community legislation should be sought along these lines. We also recommend that amendment to the list of function names of additives should be sought as should provision for manufacturers, in the absence of a specified function, to apply a descriptive function name so that the reason for all food additives used in a food is declared to the purchaser.* We note that a list of functions appears in the framework directive on food additives and that this would be a sensible starting point for discussion.

CARRY-OVER ADDITIVES

Background and legislative position

97. The Labelling of Food Regulations 1970 required the labelling of 'carry-over' preservatives and antioxidants, except in quantities insufficient to perform a technological function in the ingredient, or less than 5% of the maximum permitted amount. These were subsequently overtaken by the more general provisions in the European Community Food Labelling Directive, implemented in the Food Labelling Regulations 1984. The Regulations state that any additive whose presence in the food is due solely to the fact that it was contained in an ingredient of the food, if it serves no significant technological function in the final food, need not be named in the list of ingredients. Similarly any additive which is used solely as a processing aid need not be named. 'Processing aid' is not further defined in the legislation.

98. The question of labelling carry-over additives and processing aids was considered by the Committee between 1986 and 1988, at a time when there was growing consumer interest in additive-free foods, and a wish by consumers to be better informed about exactly what they were eating, encouraged by the policy shift away from compositional standards for foods towards more informative labelling. In addition some consumers were and remain anxious to avoid certain ingredients to which they or their families experience adverse reactions or simply because of a desire to eat 'purer' foods. Accordingly many consumers wish to be able to identify the presence of all additives whether directly or indirectly present in the final food.

Options considered and recommendations

99. Whilst neither the consultation exercise nor the consumer survey specifically addressed carry-over additives the Committee has been influenced in its consideration by the fact that both have shown that consumers generally wish to have as full and clear information about additives on the label, in order to be able to make proper judgements. We also feel that consumers generally are unaware of the subtleties of the exemptions from ingredient listing for additives, believing that the current requirements are for full, rather than partial listing, and are thus misled under the present rules. We further consider that the overriding criterion of whether the additive is serving a significant technological function in the final food is not a valid one and is not a distinction which consumers would make. We take the view that these additives may have other equally important functions in the final food beyond the technological effect, for example physiological effects, which we believe make this distinction irrelevant. *We firmly believe that carry-over additives and processing aids present in the final food at a level at which we believe consumers would wish to know of their presence, should be indicated in the list of ingredients, irrespective of their function.* However in order to balance the needs of the consumer against the significant practical difficulties which declarations of carry-over additives create for the trade, *we recommend that a de minimis level should be set for declaring the substances in the final food and suggest 10% of the maximum permitted level or 10mg per kilogram, whichever is the less. All carry-over additives, whatever their function, should be indicated in the list of ingredients subject to this cut-off level. The same principle should equally be extended to those processing aids which may appear in the final food.* However we recognise that this needs further consideration as, unlike food additives, there is no approved list of processing aids. We acknowledge that this is a difficult area and to be pragmatic *we recommend that, based on the typical levels of these substances in the final food, an indication such as 'includes x and y' at the end of the ingredients list on the food should be satisfactory.*

100. We recognised that the recommendations of a de minimis level will to a certain but very minor extent continue the situation of partial listing of ingredients which could be considered misleading. However our belief that in principle all ingredients should be declared remains unchanged. We consider that taken as a whole our recommendations to ensure ingredient listing of the constituents of all compound ingredients, to remove exemptions from ingredient listing, improve additive declarations, require listing of carry-over additives and processing aids at significant levels and to require declarations of post-harvest treatments (Section VIII) will result in a much improved, full profile of the composition of the food. We believe that it is valid for practical reasons, to adopt a different approach where substances are present adventitiously in the food.

SECTION V

PRESENTATION OF LABELLING INFORMATION

BACKGROUND

101. Ministers drew the Committee's attention to two particular aspects of presentation: whether the placing or prominence of any items of information needs to be specially controlled over and above the current requirements and whether the overall balance between essential and other information needs to be regulated, perhaps by the introduction of a standardised label. These issues were considered together since the placing or prominence of particular items clearly affects the overall balance and the starting point for both is the 'manner of marking' requirements in the Food Labelling Regulations 1984 and the practical application of these provisions. Pictorial representation and the use of symbols were also considered. As part of our assessment we looked in detail at a number of labels and were guided in particular by concern expressed by enforcement interests about the prominence of the promotional information in relation to the descriptive name of the food and other key statutory information; by support from consumer and enforcement groups for a standard panel which would contain the statutory details; by the industry's need for a flexible system and by the results of the consumer survey.

Legislative position

102. The Labelling of Food Regulations 1970 contained detailed provisions on the positioning and format of the name of the food and ingredients list, including minimum letter size requirements. These were subsequently overtaken by the more general 'manner of marking' and 'field of vision' provisions in the European Community Labelling Directive 79/112 implemented in the Food Labelling Regulations 1984. The European Commission is not expected to propose further changes in this area. The Food Labelling Regulations lay down general rules for the manner in which the compulsory labelling requirements for food are to be shown on the label. In general for prepacked foods, the labelling particulars must be shown on the packaging, or on a label attached to the packaging, or on a label clearly visible through the packaging, when sold to the ultimate consumer; certain particulars may be provided in relevant trade documents on or before delivery when the food is sold to a catering establishment. In all cases the particulars must be easy to understand, clearly legible and indelible, and when the food is sold to the ultimate consumer, they must be in a conspicuous place so as to be easily visible. They must not be hidden, obscured or interrupted by any other written or pictorial matter. The regulations require certain key information (name, quantity, minimum durability date or 'use by' date and alcoholic strength where appropriate) to appear in the same field of vision.

Views of interested parties

103. The consultation exercise showed some division of opinion among trade interests on this issue. On the one hand, the view was expressed that the present, general requirements on the manner of marking foods, if properly adhered to, should be sufficient to ensure that the essential information is given clearly and that the statutory and non-statutory information is kept apart. These respondents also pointed to the practical difficulties of having a standard label, due to the wide range of products and labels on the market, and because of the inherent inflexibility of such a system, which would undermine industry's ability to offer attractively presented foods. There were certain trade interests however, who recognised that

28

there might be advantages for the consumer in a standard panel on the label to house the essential labelling information, although again some flexibility would be needed. As regards pictorial information and symbols, the trade was generally of the view that whilst this means of presenting information had its attractions, without careful safeguards these could oversimplify the position and even misinform consumers. There was cautious support for pictorial or symbol labels in situations where they depict a particular attribute of the food, but only if sufficiently well recognised and not as a substitute for proper written information.

104. Enforcement interests thought that the impact of the essential statutory information on labels was being reduced by the increasing amount of promotional information given on labels. They therefore favoured standardising the presentation of the essential information by means of a statutory information box, to be displayed in the most prominent position on the package. Enforcement interests were also concerned about the lesser prominence given to the descriptive name as opposed to the trade, brand or fancy name. Minimum letter size requirements and controls on colour contrast were suggested. As regards pictorial and symbol labels there was general consensus that, in certain circumstances, these could be useful for consumers and that they should therefore be permitted, but not encouraged. On the whole however, it was felt that these were not an acceptable substitute for clear and precise written information.

105. There was concern among consumer groups that non-essential promotional material often took precedence over the statutory information and cautious support for some form of standardisation of the label, or for a standard panel on the label for the essential information. Many respondents however felt that consumer market research was necessary to determine whether consumers would, in fact, consider this a positive improvement. There was also recognition of the practical difficulties for the trade which may be associated with standardisation of information. A few respondents proposed minimum letter size controls to ensure easily readable information. There was considerable support from consumer organisations for information in pictorial or symbol form, which they felt might be simpler and easier to understand. The suggestion was made that this means of labelling should be considered as an alternative to an increase in the requirements for statutory information. Several respondents urged caution in respect of pictorial or symbol labels until further evidence of their efficacy was obtained, since unless symbols were well-recognised and understood, consumers might be confused rather than informed. On the whole this means of labelling was seen by consumer groups to have a useful place, but perhaps in a limited capacity and probably non-statutory.

Consumer survey

106. 64% of respondents in the consumer survey thought that the information given on food labels was in general clearly presented (i.e., easy to read and not hidden away). Of those who thought that generally the information was not clearly presented, the majority (52%) said this was because the print was too small. A wide variety of other reasons, all spontaneous, were given by the remaining 48% (such as price tag obscuring important information, information lost through lack of colour contrast) but each reason was put forward by no more than 7% of respondents. Those respondents who considered that labelling information was not clearly presented were also asked in what ways they thought presentation could be improved. Again a significant proportion (44%) suggested larger or clearer print, and other improvements were suggested by only 10% or fewer respondents; 7% sought standard placing. It was also interesting to note in this context that 64% of respondents disagreed or disagreed strongly with the statement that 'too much information appears on food labels'.

Options considered and recommendations

107. We recognise and have sympathy with the concern expressed by many respondents that the non-statutory information on food labels is tending to swamp the essential statutory details and undermine the balance which should exist between them. In our view improvements to the present general requirements for the manner of marking are needed to ensure that consumers are able to obtain the essential information quickly and easily. We considered whether more specific rules, for example minimum letter sizes, would help to increase the prominence of key items and right the balance, but we rejected this option since it seemed unnecessarily restrictive without obviously being effective. In particular it would be possible to adhere to letter size requirements but make the letters inconspicuous through use of colour and print types. Any new regulations would therefore have to be very detailed to ensure proper control.

108. We looked carefully at the suggestion by enforcement and consumer interests that a standard panel, containing all or some of the statutory labelling information, should be introduced. We were impressed by the clarity of certain labels, such as those for pet foods, using a statutory box format for key information and *we concluded that a standard panel, prominently displayed on the label, would help to counter the difficulties over the relative prominence and placing of the statutory and non-statutory information.* We also considered what information should be contained within the panel; whether it should be limited to 'food safety' information such as the date-mark, storage instructions and instructions/conditions for use or whether it should be more comprehensive. Whilst we recognised the dangers of overloading the standard panel the Committee's view was that *all the statutory information, (with the exception of the name and address of the responsible manufacturer, packer or seller), together with nutrition labelling information and any additional safety information should be included in the box. All other information should be excluded from the box.* However in order to give some flexibility to the requirement, particularly for small labels, *we recommend that 'signposting' (for example 'see lid', 'on back of pack') as currently permitted for the date-mark should also be allowed for the following information:*

    (a)   storage instructions;

    (b)   conditions of use;

    (c)   instructions for use;

    (d)   nutrition labelling;

We recommend that the following statutory requirements should always be given in the panel:

    (a)   name of the food;

    (b)   ingredients list;

    (c)   weight;

    (d)   alcoholic strength (if appropriate)

    (e)   place of origin, if its omission might mislead;

109. We decided against making specific recommendations concerning the positioning and prominence of the panel since we believe that the current rules on the intelligibility of the statutory particulars fully cover this. However we would draw industry's attention to these requirements in the Regulations, in particular that the information 'shall be marked in a conspicuous place in such a way as to be easily visible' and 'not in any way be hidden, obscured or interrupted by any other written or pictorial matter. *We would encourage industry to respond to the very clear concerns about the size of lettering used on food labels, as indicated by the consumer survey, and to take account of the difficulties experienced by older people and others with impaired vision.* 'Signposting' of certain details within the standard panel should allow all the information to be clearly legible.

110. As part of its consideration of the overall balance of the label the Committee also considered whether further controls were desirable for the non-statutory promotional material on labels, including illustrations. However we concluded that the statutory standard panel, recommended above, would provide an effective counterbalance to this material. We do not wish to prevent manufacturers from marketing their products well or detract from the enjoyment which attractively presented products can give. Nevertheless we would remind industry that all the details and pictures given on food labels and in advertisements must be such that they would be unlikely to mislead consumers and that the Food Safety Act 1990 also makes it an offence to present food in a way which is likely to mislead as to its nature, substance or quality. 'Presentation' has a wide interpretation in this context.

111. The Committee looked at the difficulties respondents had identified concerning the relative prominence given to brand and fancy names over the true or descriptive name of the food and considered whether further rules were needed, in particular whether the true name should be required to be as bold and compelling as the brand or fancy name and no less conspicuous than any other information on the label. However, whilst recognising the merits of this suggestion, we felt that it would considerably reduce the flexibility for label design without necessarily increasing consumer protection, particularly as many true names, in order to be fully accurate, are relatively long and detailed and we would not wish to discourage this. The name of the food must, of course, comply with the intelligibility provisions of the Regulations which require it to be clearly legible and in a conspicuous place in such a way as to be easily visible. *We do, however, recommend that the true name should be required to appear in immediate proximity to the trade or fancy name most prominent under normal conditions of purchase in addition to its inclusion in the standard panel as recommended above.*

112. Finally we considered whether the overall balance of the label and clarity of information could be improved by the wider use of pictorial or symbol labelling which is being used effectively on a voluntary basis to convey important information to certain groups of consumers, for example vegetarians and coeliacs. We welcome the fact that the Ministry of Agriculture, Fisheries and Food is sponsoring research into graphical representation of nutrition information since we believe the use of symbols in this area may be a valuable help to consumers in addition to the numerical information. However, whilst our view is that pictorial or symbol labelling may be useful in certain circumstances and should therefore be permitted, we agree with respondents to the consultation exercise that such labelling should not be a replacement for clear and precise written information. we therefore make no recommendations in this area for the present and believe that any further action should await the results of the Ministry's nutrition labelling research.

SECTION VI

CLAIMS

BACKGROUND

113. We considered, in addition to the advice we have already offered on nutrition claims and the use of the word 'natural', whether the present and forthcoming statutory requirements represent the essential controls required to protect and inform consumers in relation to claims and whether additional controls on food claims on labels and in advertisements should be introduced beyond those already in place. We looked at the general principles that should apply to all claims and gave detailed consideration particularly to the area of medicinal and health claims, including endorsement schemes and testimonials.

114. The borderline between medicinal claims, which are subject to strict controls, and health claims, which are not subject to specific controls, is unclear and the increasing use of health claims in the labelling and advertising of certain foodstuffs has led to a great deal of controversy. This has occurred both here and in the USA and recent market research, including the consumer survey, indicates that consumers want more controls over health claims and that current legislation may be too imprecise. Concern has been expressed that some manufacturers may be exploiting the current interest in health, while the public is unsure of the validity of the claims. This has led to a need to examine and clarify current UK legislation, particularly as preliminary draft European Commission proposals covering these claims have been issued.

Legislative position

115. The existing framework of law in the Food Safety Act 1990, the Trade Descriptions Act 1968 and the Food Labelling Regulations 1984 provide controls over misleading descriptions and claims in food labelling and advertising. The Food Labelling Regulations contain detailed controls on certain specific claims, both express or implied, made in the labelling or advertising of foods. They ban tonic claims, and claims that baby foods are equivalent or superior to the milk of a healthy mother, and set out controls on claims for foods for babies and young children, foods for particular nutritional uses, foods for diabetics and 'slimming' foods. Medicinal, vitamin and mineral, polyunsaturated fatty acid, cholesterol and energy claims are also controlled, as are claims which depend on another food.

116. It is expected that changes will become necessary following agreement on the latest European Community Framework Directive on Foods for Particular Nutritional Uses (PARNUTS) which was adopted in May 1989 and which must come into operation by May 1991. The Directive largely repeats the 1977 Directive which it replaces,but it removes the scope for national derogations from the European Community rules and adds a 'safeguard' clause which provides for a simple notification system for newly marketed PARNUTS products. The main changes to the UK's claims regulations will, however, be in relation to certain categories of food, identified in the Framework Directive, which will be subject to the provisions of nine detailed directives covering infant formulae, follow-up milk and other follow-up baby foods, low energy and energy reduced foods for weight control, dietary foods for special medical purposes, low sodium foods, gluten-free foods, foods for sportsmen and foods for diabetics. It is intended that these directives will regulate both the composition of the foods and their labelling, presentation and advertising, including permitted claims.

117. A recently produced preliminary working paper referred to in paragraph 114 above, sets out the Commission's first ideas for a future European Community Directive on claims. We have taken this into account in making our recommendations.

Views of interested parties

118. The majority of respondents did not comment on the control of claims, but those who did emphasised that consumers needed to be properly informed and not misled. Particular concern was expressed over the use of the term 'light' which was ambiguous if unqualified and was confusing to the consumer who could not reliably determine the meaning of the claim. There was also felt to be a need to clarify the law in relation to health claims so as to ensure that only truthful and scientifically substantiated claims were permitted. This view is in line with the concerns outlined in paragraph 114 above and with the findings of the consumer survey.

Consumer survey

119. The survey did not address this area in any detail. However more than half (56%) of respondents agreed with the statement 'I find it difficult to believe some of the health messages appearing on some products nowadays'. Less than a quarter (23%) disagreed.

USE OF THE WORD 'NATURAL' AND NUTRITION CLAIMS

120. Comments were received from respondents in relation to the use of the word 'natural' and nutrition claims which refer to the content of nutrients in the food. However our advice on these claims was issued recently and we have not reviewed it as we believe it holds good. The Committee's guidelines on the use of the word 'natural' are attached at Appendix IV and we understand that the Ministry of Agriculture, Fisheries and Food is looking at uptake of the guidelines to see whether further action is necessary pending European Community legislation. Our report recommending legislative controls for certain nutrition claims is attached at Appendix V. These recommendations have been accepted by the Government but legislation has been delayed because of the European Community's intention to make proposals in this area.

GENERAL PRINCIPLES

121. In considering a possible framework for control, within which claims would be allowed, we were aware that the current UK Regulations seem to have developed on an essentially piecemeal basis. When we considered the need for controls over the term 'natural' and on nutrition claims our recommendations were made in the light of a series of principles which we believe provide a clearer and more cohesive approach to regulations. we therefore considered whether this approach could be extended and identified *the main principles,* which we believe should apply to all claims expressed in generally applicable terms, as:

(a) a food must be able to fulfil the claim being made for it and adequate labelling information must be given to show consumers that the claim is justified;

(b) where the claim is potentially ambiguous or imprecise it must be clearly explained on the label as well as justified (see paragraph 123 below);

and, from the 'natural' recommendations,

(c)   a claim that a food is 'free from' a substance or treatment should not be made if all the same class or category of foods are similarly free from 'x' (though we recognised that some statements of this type might provide accurate and beneficial information for consumers);

(d)   words or phrases which imply that a food is free from any specific characteristic ingredient or substance should not be used if the food contains other ingredients or substances with the same characteristic. This is intended to prevent claims such as 'free from preservatives' when ingredients such as vinegar have been added for their preservative effect. (A claim such as 'contains no animal fats' would not be caught if these had been replaced with non-animal fat as in this case the key characteristic is the animal origin, not the fat);

(e)   meaningless descriptions (such as the use of 'special', 'selected', 'healthy', 'wholesome' without further explanation) should not be used;

and, from the nutrition claims recommendations,

(f)   comparative claims must be justified against relative and generally applicable criteria (for example 'reduced' may only be claimed where the reduction in the claimed nutrient is 25% or more);

(g)   the label should give a sufficiently full description of the food in relation to the area for which the claim is made to ensure that selective claims, even if true, do not mislead;

(h)   absolute claims must be justified against absolute criteria set for a given nutrient and applying to all foods (for example 'low fat' may only be claimed when the total fat content is not more than 5g per 100g and per serving);

(i)   where a food is naturally 'low' or 'high' in a substance the claim should be 'a low/high x food'.

The wider application of general principles appears to be in harmony with the European Commission's initial thinking on claims and we consider that this concept offers a possible solution to a difficult legislative problem given its potential advantage of providing a more readily understandable framework which could be of benefit to manufacturers, enforcement authorities and consumers.

AMBIGUOUS CLAIMS

122. At present there is no overall control over claims which are ambiguous and in particular the description 'light' or 'lite' is being used on a wide range of products (including alcoholic and non-alcoholic drinks, yoghurt, and reduced fat spreads) and in potentially confusing ways. Often the description is used to imply that the product is a 'healthier' version of a standard one but it is frequently unclear just how the traditional formula has been adjusted (for example reduced alcohol, fat, sugars, calories) or even if the formula has been changed (for example 'light lager' may simply refer to a light taste). To add further to the confusion the description is also used for traditional products such as 'light ale' and 'light brown sugar'. Similarly, the description 'diet' or 'diat' is often used in connection with alcoholic or soft drinks e.g. 'Diat pils', 'Diet cola' and in some cases the meaning of the description is unclear. This applies, in particular, to the kind of imported German lagers where the description 'diat' in fact simply refers to the beer emanating from a particular area in Germany and not to an indication that it is low in calories or alcohol.

123. The regulations relating to slimming claims would have to be met if it was claimed that a 'light' or 'diet' product was an aid to slimming or weight control, and it is possible that some fat-rich products claiming to be 'light' might need to conform to our recommendations on implied comparative nutrition claims. However we consider that compliance with the general principles outlined above would ensure that consumers were adequately informed. The basic aims of the existing controls in the Food Labelling Regulations are to ensure that the food can fulfil the claim (direct or implied) being made for it and that the information showing how the food achieves this is given on the label. These aims are also embodied in the European Commission's working document on claims, which would only allow claims relating to measurable and objective characteristics and then only if they could be substantiated. *We consider it important that any potentially ambiguous or imprecise claim made in the labelling or advertising of a food must be explained and justified on the label, and be capable of substantiation. We therefore believe that the use of terms such as 'light' and 'diet' carry the obligation for the product label to include an additional description explaining the context(s) in which the claim is used (reduced alcohol, low calories, light colour etc) and, where appropriate, quantifying the claim.*

MEDICINAL AND HEALTH CLAIMS

124. There is a general prohibition, throughout the European Community, on claims which attribute to any foods, except PARNUTS foods, the property of preventing, treating, or curing a human disease. Disease includes any injury, ailment or adverse condition, whether of body or mind. At present the UK may permit such claims for PARNUTS foods but once the Framework Directive referred to in paragraph 116 has been fully implemented, medicinal claims will only be permitted if they have been approved under European Community procedures. The general ban on medicinal claims is contained in the Food Labelling Regulations 1984 which state that a claim that a food is capable of preventing, treating, or curing, human disease may only be made if a product licence has been issued under the Medicines Act 1968. Although the Act itself relates only to products which are wholly or mainly used for medicinal purposes, which means that licences are not issued for products which are mainly foods, it is likely that a PARNUTS food which warranted a medicinal claim would be considered to be mainly medicinal in its purpose and would therefore be covered.

125. In view of this, it seems that the controls on medicinal claims, which are effectively banned for products which are mainly foods, is in line with Community obligations and *we recommend that the present law regarding direct medicinal claims should be maintained*. This view is strongly supported by consumer and health groups, some of whom would also like to see the law clarified for implied medicinal claims which overlap with health claims.

126. It is at present uncertain which health claims are permitted because the Food Labelling Regulations are not explicit as they directly ban only absolute medicinal claims (for example 'this food is capable of preventing heart disease'). If a qualification is inserted (for example 'this food may help reduce the risk of heart disease') a view has to be formed whether an implied claim, covered by the Regulations, has been made. In practice health claims distance themselves still further from the clearly prohibited claims and state for example, that a product will 'help reduce cholesterol levels when eaten as part of a low fat diet' or that it will 'help maintain a healthy heart'. There is no mention of disease and whether the claim breaches the Regulations is a matter of opinion since the dividing line between an implied medicinal claim and a health message is far from clear. We have strong reservations about the value of such claims, whether or not they can be fully justified and substantiated. Claims which have little substantiation are likely to be misleading and to undermine consumers' confidence in labelling and health education information. In the limited space available on the label it is difficult to put the claim into its proper context and qualifications such as 'as part of a low fat diet' are easily lost.

127. In considering this issue we took account of developments in the USA, where the Food and Drugs Administration (FDA) is seeking to reach a regulatory position to allow health claims, when they are appropriate, whilst still protecting the public against fraudulent or misleading claims. We were aware that the FDA considers that claims regarding the role of food in the prevention, cure, mitigation or treatment of a disease are essentially medicinal in their nature but that claims regarding the effect of food on the body 'need not make the food a drug if the claims relate to how the food affects the structure and function of the body'. Thus, for example, an indication of the role of calcium in building strong bones and teeth would generally not be a 'drug' claim. It is stated that foods have such effects by virtue of their nutritional value when consumed over time and not as a result of an immediate pharmacological response, as is the case with medicines. We felt that this view had merit.

Claims related to deficiency diseases

128. We recognise that the nutritive value of a food includes the usefulness of a food component in reducing the risk of a disease or condition which is caused solely as a consequence of that component being deficient in the diet. However, we consider that a claim relating to a deficiency disease which is virtually non-existent in the European Community could be misleading as it could convey the impression that an ordinary diet is not sufficient to prevent the disease. We *recommend that such claims should be prohibited,* unless it is made clear that the product is aimed at a particular minority with special needs (for example in the case of some vitamin and mineral pills which are classified as foods under UK Food Law).

Claims related to chronic diseases

129. We believe that a food's nutritional value may also include the usefulness of a food component, consumed as part of a total diet, in reducing the risk or forestalling the premature onset of a chronic disease condition. In the light of this, we considered where there is appropriate scientific evidence, whether it is appropriate to allow claims which set out the perceived benefits of a food (such as for a food low in saturated fat in relation to cardiovascular disease) as this can be useful to consumers who desire to adopt a healthier dietary pattern. The FDA has indicated that it is considering permitting these claims, subject possibly to a prior approval system. Given that the USA and the UK have had similar experiences of health/medicinal marketing campaigns which have caused concern, there seems an advantage in having a coherent, justifiable and enforceable policy of control which ideally could be applied to both sides of the Atlantic. Against this the preliminary draft European Community proposals appear to be moving in the opposite direction, and seek to prohibit claims which refer to the connection between nutrition and health unless authorised by European Community legislation. We have some sympathy with this restrictive approach because of the potential for abuse. We noted, however, that there is no presupposition that health claims will be completely banned in the European Community and we looked to see if a balance could be struck which would continue to provide for the possibility of allowing a very limited number of health claims to be made, provided that the dividing line between these and medicinal messages could be clarified.

130. It is arguable that there is merit in using food labels to present scientifically well-founded health messages provided that the message does not mislead or deceive the consumer, and provided that the legal position prevents abuse. Advocates of this view point out that consumers have the right to be told of a food's particular health benefits and that denying this information would inhibit their ability to make informed choices and would tend to limit the development of 'healthier' food. We are told that health messages do influence purchases, whether these messages appear on the label or become known through the media. On the other hand, it is argued that, whilst the whole diet is made up of individual foods all of which have a part to play, most individual foods have a small impact on the overall diet and thus consumers are unlikely to derive significant benefit from consuming a normal quantity of the food. Furthermore an abnormal quantity of the food might work against achieving a balanced diet. We share these views and believe that in general health claims do *not* have a place in food labelling and advertising. However we believe that given strict controls, certain health claims, the scientific basis of which has been endorsed by a specified authoritative body, can provide useful information.

DEFINITION OF 'HEALTH CLAIM'

131. Before reviewing possible means of control we considered what may be defined as a health claim and concluded that this is 'any statement, suggestion or implication in food labelling or advertising (including brand names and pictures) that a food is in some way beneficial to health, and occurring in a spectrum between, but not including, nutrient claims (e.g. low fat, low cholesterol, high fibre) and medicinal claims'. Within this range health claims may be broken down into three broad types:

(i) where a reference is made to a possible disease risk factor (e.g. can help lower blood cholesterol);

(ii) where an aspect of a nutrient's effect on the body is explained (for example the role of calcium in building strong bones and teeth);

(iii) general health claims outside these categories (for example that a product is a healthy food).

OPTIONS FOR CONTROL

132. We considered the possible means to control health claims but concluded that a 'prior approval' system (whereby no food would be able to carry a health claim unless it had been pre-approved) and/or a system of approving health messages for general use, would be unnecessarily burdensome both to adminster and to comply with. A unilateral system would not be possible given that food labelling is harmonised throughout the European Community.

133. We also looked at the type of 'safeguard' system to be used in the case of some PARNUTS foods to see if it might be extended to require the labels of all newly marketed foods making health claims to be sent to the relevant food authority which could seek information from the company to verify the claim. Given that many PARNUTS foods are products for which justifiable health/medicinal claims are likely to be made there seemed to be a certain logic in extending the principles already agreed for these foods at Community level to other non-PARNUTS foods for which such claims may be made. Against this we considered the expense that would be incurred by companies in making up an appropriate dossier and the likely difficulties this would create for small companies which wished to initiate claims or to copy other manufacturers' claims. We also acknowledge that additional resources would probably be required to administer any of these systems although these might be partially offset by the benefits of having readily understandable and enforceable laws. Overall we concluded that the potential benefits of a 'safeguard' system were insufficient to justify its imposition.

134. We concluded that it would be better to set strict criteria against which any permitted health claims could be assessed. In particular *we believe that health claims should only be permitted if they can be justified in relation to any recommendations that have been made or supported by the Chief Medical Officer (CMO)*. The recommendations of similar Government advisers or expert bodies in other countries and other scientific opinions would be an acceptable basis for a claim if these recommendations have been supported by the CMO. This would not be a prior approval system and claims would therefore only be permitted in areas which the CMO and his expert advisory committees had addressed. *We recognise that this would be stringent, and would effectively ban 'novel' claims until the scientific evidence had been considered by the CMO, but firmly believe that the present potential for making misleading or spurious claims justifies this approach.* Ideally we would like to see an expert body established specifically to advise on the validity of health claims on food labels and advertisements and to police the system. However we recognise that this may not be possible hence our recommendation that the CMO's advice should be the basis for claims. We nevertheless acknowledge that if our recommendation is accepted it might significantly increase the workload of the CMO's expert advisory committees. In making this recommendation we were also aware that the preliminary draft European Community claims proposals would ban references to recommendations by the medical profession or nutrition or health bodies except in accordance with European Community provisions, as yet unformulated. *We do not think this is helpful and would wish to allow manufacturers making permitted health claims to refer to recommendations by the CMO, where appropriate, to help substantiate their claims.*

135. In addition, we considered what principles might be used to control permitted health claims and recommend that:

(a) *the claim must relate to the food as eaten rather than to the generic properties of any of the ingredients;*

(b) *a food (when consumed in normal dietary quantities, (see below) must be able to fulfil the claim being made for it and adequate labelling information must be given to show consumers that the claim is justified;*

(c) *the label should give a full description of the food to ensure that selective claims, even if true, do not mislead and any claim should trigger full nutrition labelling (at least the group of eight nutrients in the EC Nutrition Labelling Rules Directive);*

(d) *the role of the specific food should be explained in relation to the overall diet and other factors.* This may require just a simple statement, such as 'if eaten as part of a low fat diet'.

The insertion in (b) above, of the phrase 'when consumed in normal dietary amounts' has been included as we feel that claims should only be made for foods which can reasonably be expected to make a significant contribution to a balanced diet. Whilst this would obviously rule out claims for foods likely to be consumed only occasionally or in small quantities (such as condiments) we recognise it would leave a grey area where judgement would need to be used. For example, it might need to be decided if only the food which was subject to the actual claim should be considered, or if it was legitimate to consider the likely accumulated beneficial effect of consuming various other similar foods (eg reduced fat cheese alone or all reduced fat products).

FOOD HEALTH ENDORSEMENT SCHEMES

136. Another aspect of health claims we considered was the use of food endorsement schemes which promote certain types of food (eg pasta, bread, lower fat spreads) as being beneficial to health. At present the main scheme of this type is run by the Health Education Authority (HEA) and was established in the belief that food labels and product advertisements can effectively be used to convey justifiable and useful health messages to consumers, though it is acknowledged that safeguards are needed to prevent abuse.

137. We found this to be a grey and difficult area involving subjective judgement and before deciding on our recommendations we exceptionally decided to request representatives of the HEA to attend one of our meetings to explain the background to their scheme and how it was run. We were told the Authority's policy and practice of endorsing certain foods had been reviewed by the Department of Health, in collaboration with the Ministry of Agriculture, Fisheries and Food, and it had been concluded that there was a place in nutrition education for positive endorsement of this type and that, in order to encourage changes which would lead to people eating better balanced diets, such endorsements were most effective on product labels or advertisements in association with brand names. In mounting such a scheme the HEA sought to ensure that COMA and other Government recommendations (including those from our Committee) were strictly followed and these form part of their criteria for deciding which types of food are potentially endorsable. The HEA aimed to be absolutely scrupulous by offering opportunities equally to all producers of 'endorsable foods'. This was considered to be an acceptable compromise between the need for authoritative and impartial advice on nutrition – clear of any commercial interests – and the need to make progress in nutrition education and in achieving dietary changes, in particular to help reduce the incidence of heart disease. We understand that the Authority's decision to work with commercial interests was not taken lightly but was made in order to encourage improvements in the national diet, as recommended by COMA for public health reasons, since educational campaigns alone had hitherto resulted in few actual changes. The HEA view is that consumers react positively to their endorsement schemes.

138. Whilst accepting that the HEA's approach was a pragmatic response to the need to improve the national diet and that it was even-handed and based on sound dietary criteria, we were not convinced that the benefits of endorsements outweighed the likely drawbacks. We therefore considered whether their system could form the basis of a tightly controlled model within which endorsement schemes could be permitted. However, our fundamental concerns remained. We accepted that there were some advantages in promoting foods which would contribute positively to a balanced, healthy diet but we felt that endorsements on labels and product advertising were not the way to do this. We believe that food labels are primarily for informing consumers about the individual product. Their potential for educating consumers is necessarily limited by the space available and by the circumstances in which purchases are made. In our view it is a practical impossibility for the overall dietary message to be promoted on food labels. We accept that the Authority has tried to overcome this by making their food endorsement activities part of specific campaigns with attractive, authoritative and accessible educational material. However we do not feel that the back-up of an educational campaign is sufficient to overcome our reservations. Indeed we believe that label endorsement might assume more importance for the consumer than the educational material.

139. Endorsement schemes require significant resources to operate properly, both to ensure good science and the tight policing which we believe are necessary to ensure conformity with scheme conditions and to prevent unjustified exploitation by sponsors or by the industry. The aims of food endorsements, notably to bring about dietary changes, are not primarily matters for food labelling; neither are they appropriate for product advertisements. We believe that the risk of misleading consumers is very great in this area and that it would be difficult to control health endorsements in the same way as health claims. Given that there is no suitable existing mechanism for introducing controls we feel it is important to err on the side of caution. *We recommend that health endorsement schemes should not be permitted and that the Government considers other ways to achieve dietary change by improving basic nutrition education and explaining its relationship with health.*

HEALTH ENDORSEMENT BY INDIVIDUALS (TESTIMONIALS)

140. We also looked at the question of endorsement by individuals on food labels, or in advertisements, in ways which amount to a health claim because they refer directly or implicitly to the health-giving properties of the product. Such testimonials can be differentiated from the food endorsement schemes referred to above since they are characterised by stated or implied product claims which seek to acquire conviction because of their association with an individual, usually famous, consumer. We see no problem in principle with promotion by famous personalities or sponsorship of events which may be designed to give the food a 'healthy' image. We are concerned that consumers may be seriously misled by explicit or implied claims making reference to possible disease risk factors or more generally to what the food can do to improve or maintain health. Another facet of testimonials is the promotion of products by so-called 'experts'. We recognise that action can already be taken against any misleading promotions under the Food Safety Act 1990 and the Food Labelling Regulations 1984, which apply to both labelling and advertising, and that they would also be subject to any amended legislation which may be introduced with our recommendations on health claims. Nonetheless we feel that testimonials amounting to health endorsements can significantly influence consumer choice and that there is considerable scope for misleading consumers, particularly as there is no way for consumers to assess the information or establish the bona fides of the 'experts'. *We therefore firmly recommend that all such health endorsements be banned.*

SECTION VII

METHODS OF REARING AND SLAUGHTERING

BACKGROUND

141. The Committee was asked to address the question of whether particular methods of rearing or slaughtering animals or poultry should be required to be indicated on food labels. This arose from requests to Ministers for additional labelling requirements in this area, particularly concerning methods of rearing and primarily for ethical reasons since animal welfare is of considerable interest and importance to many people. Given this background, to help in our consideration of the need for and practicality of any new requirements, we looked in some detail at current animal welfare policy.

142. It is an offence under the Agriculture (Miscellaneous Provisions) Act 1968 to cause livestock on agricultural land any unnecessary pain or unnecessary distress. Regulations have been made under the Act some of which affect husbandry systems directly. The Welfare of Calves Regulations 1987 effectively ban the traditional veal crate system. The Welfare of Battery Hens Regulations 1987 implement the European Community Directive on battery hens, and inter alia, lay down minimum space allowances for caged birds. In due course, further standards are expected to be introduced in the European Community for production systems for the main livestock species. European Commission proposals on pigs and calves have been submitted to the Council of Ministers and are currently under negotiation.

143. A major plank of animal welfare policy in this country is the Welfare Codes made under the Act. These Codes are based on recommendations from the Farm Animal Welfare Council (FAWC), the Government's independent advisory body, and have the approval of Parliament. Codes now exist for pigs, sheep, cattle, domestic fowls, turkeys, ducks, rabbits, farmed deer and goats. They lay down basic recommendations necessary to protect the welfare of farm livestock. They are not mandatory, but where a person is charged with causing unnecessary pain or distress, a breach of a Code recommendation may be used in evidence and will tend to establish the guilt of the accused. The key factor in safeguarding the welfare of livestock is the level of stockmanship and all of the Welfare Codes refer to this in their introductory passages. It is recognised that while the husbandry system is important when considering the welfare of the animals, it is possible that a 'welfare friendly' system under bad stockmanship might lead to poorer welfare for the livestock than an intensive system with good stockmanship. The system alone does not necessarily dictate the welfare status of the animals and there is thus potential for consumers to be misled by labelling which indicates methods of rearing.

144. In addition to the views of interested parties and the results of the consumer survey we had the benefit of the views of the FAWC which advocated positive labelling to identify welfare-orientated systems and suggested that a single standard system of ethical labelling would ensure that producers were fairly treated and that consumers could make choices according to their convictions.

145. As far as labelling of slaughtering methods is concerned we focused in particular on the question of slaughter by religious methods since this alone was specifically mentioned by respondents to the consultation exercise. We were aware that the FAWC, in its 1985 Report on the Welfare of Livestock Slaughtered by Religious Method, recommended labelling of carcases and cuts to indicate method of slaughter and of the reasons why that recommendation was not accepted.

Legislative position

146. It is not a principle of general UK or European Community food labelling legislation that information about methods of rearing or slaughtering animals or poultry should be required on food labels. Labelling requirements are designed to give full information about the final product and not about the stages prior to it becoming a food. Voluntary labelling indications are permitted subject to the general provisions of the Food Safety Act 1990 and the Trade Descriptions Act 1968 which prohibit false or misleading descriptions of food.

147. The hen's egg is the only example of animal produce where specific labelling rules relating to the method of production are in force. Indications of the production method on egg packs are tightly controlled by European Community regulations which limit the terms that may be used (to free-range, semi-intensive, perchery (barn and deep litter) and lay down minimum specifications that have to be met before such terms may be used. A European Community regulation on poultrymeat marketing standards, which has recently been adopted, provides for the European Commission to regulate the definition of indications of the type of farming for poultrymeat covered by the standards. It is expected that the Commission will in due course present proposals which will cover non-intensive methods of farming.

Views of interested parties

148. Trade interests consulted believed that it would be impracticable to undertake detailed labelling of this kind in view of the diversity of supply of such products and that it would be unreasonably burdensome and expensive to achieve. Some suggested that effective enforcement would be virtually impossible, particularly if any requirements were extended to compound meat products. The trade view was that statutory labelling should be restricted to aspects affecting the quality of the final food. Reservations were also expressed about the difficulties of describing the whole range of rearing and slaughtering methods in a consistent, meaningful and objective way for consumers. The difficulties of ensuring satisfactory compliance with any given definitions, especially taking into account imported foods, were also mentioned. Many trade respondents proposed that sustainable claims about particular methods of rearing or slaughtering should continue to be permitted on a voluntary basis as this system was more flexible and responsive to the needs of individual groups or consumer trends than a blanket requirement.

149. Some enforcement interests said that many consumers wanted information and that freedom of choice should therefore be respected, despite the practical difficulties which might be involved. Others pointed out that the trade already responded to such demands where there was a commercial advantage. In common with the trade it was argued that information of this kind should not be required where the quality of the final product was unaffected. There was recognition of the practical difficulties which statutory labelling requirements would involve and enforcement interests pointed to the need for carefully defined terms to describe the various production methods if consistency and meaningful labelling were to be ensured and abuse prevented.

150. There was no clear view from consumers on this particular question. On the one hand some consumer interests took the view that it was an impracticable proposition, questioned whether the food label was the most appropriate means of conveying this information to the consumer, and for this reason suggested it should remain on a voluntary basis. The practical difficulties of effective enforcement were also mentioned. Other consumer groups were cautiously supportive of this form of labelling but pointed out that the terms would need to be specified in simple language and carefully defined so that consumers were not misled. It was proposed that this should preferably be achieved at European Community level to ensure consistency and prevent trade distortions. Some consumer respondents suggest that it would be impossible to arrive at simple terms and definitions which could, for example, take account of both form of husbandry management and the method of slaughter in a meaningful way and that consumer confusion was the more likely outcome.

Consumer survey

151. The intention of one of the questions of the consumer survey was to establish which items, out of eleven that could be required to be given on labels in the future, would respondents really use and refer to. This was a prompted question. The type of farming methods used and slaughtering methods were two of the eleven items. Respondents were asked to sort the eleven items into those they thought they would use and those they would not, bearing in mind the limited space available on food labels. The results show that 40% of all respondents indicated that they would refer to and use the information on farming methods, placing this item about halfway on the list. Almost one third of those who said they would use this information gave concern about 'unnatural'/intensive farming practices as their reason. 24% specifically mentioned battery methods of chicken and egg production in this context. It should be noted however that a slightly larger percentage (42%) indicated that they would refer to and use information on which foods were organically produced. Quantities of the main ingredients in foods, whether pesticides were used, and more detail on general descriptions all came higher on the list of priorities. Overall, respondents showed least interest in referring to and using information about slaughtering methods (19%). Slaughtering methods were however of particular concern to those from ethnic minorities, 61% of whom said they would use this information. These results should be considered against the response to a prior question in which 75% of all respondents said that there was not any information which should appear on food labels that currently does not. Of those who thought information was lacking, none (in an unprompted situation) mentioned farming or slaughtering methods as information which was missing.

Conclusions and recommendations

152. In our consideration we took particular note of the fact that the consultation exercise and the consumer survey both indicated that a substantial proportion, although probably a minority, of consumers wish to have additional labelling to indicate farming/production methods for foods and that ethical considerations are behind the request, with concern about intensive farming practices being mentioned most frequently. There would not seem to be any concern over the quality of the final food produced by the various systems.

153. All interests pointed to the considerable practical difficulties which would be associated with a general requirement for such labelling and we share this view. There would be a need to set standards for the various systems to be indicated so that they could be clearly defined but even then the large number of systems involved and the fact that more than one system may be used in producing the final food would create considerable problems for labelling. Moreover it would not be possible to take account of the stockmanship which is critical to the welfare of the animal and there is thus potential for consumers to be misled. We believe that there is significant room for consumer confusion unless the labelling terms for the various systems can be carefully defined and specified.

154. As well as the difficulties in drawing up a labelling scheme that was clear and meaningful, we also recognise the difficulties that such a requirement would create for industry and enforcement officers since physically and analytically there is no difference between the products produced under the different rearing systems. To meet a statutory requirement new record keeping systems would have to be introduced with documentation throughout the chain from producer to retailer. This would be a severe burden resulting in increased costs which would probably be passed onto the consumer.

155. It remains arguable whether a labelling requirement is impracticable but in our view some of the difficulties mentioned above are not insurmountable. Whilst we took practicalities into account, this was not the key factor in our conclusion. We fully recognise consumers' concerns about animal welfare but we do not believe that labelling should be seen as the answer to these concerns and nor should it be used as a means of encouraging improvement in welfare standards. Animal welfare is a citizen issue rather than a consumer issue and high standards should be sought by direct means not indirectly through food labelling. Information to consumers about matters of animal welfare should be conveyed by other means such as education. Our view is that labelling requirements are to enable informed choices to be made about the final food itself and since there is no difference in substance or quality between foods produced from animals reared under various husbandry systems *we do not recommend further requirements to indicate animal rearing methods on food labels.*

156. However we acknowledge that some sectors of the trade are already responding to consumer demand for information in this area and that this may increase. *We are concerned that consumers should not be confused or misled by the terms used and we therefore recommend that further consideration should be given to whether it is possible to draw up a standard system with clearly defined terms to indicate welfare-orientated rearing methods which would help control this type of claim.*

157. The Committee also took note of consumers' slightly greater interest in organic labelling, as indicated by the survey, and of the progress being made on European Community legislation on organic foods. These proposals which include production standards and labelling provisions are expected to be adopted shortly but they do not at this stage cover animal products. *We would urge extension of the regulations to animal products as quickly as possible and that rearing methods should be fully taken into account.*

158. On indicating methods of slaughter, the consultation exercise and the consumer survey both showed that most consumers place this far down the list of priorities for additional labelling, although a high proportion of those from ethnic minorities requested this information, the main reason being a positive need to identify the meat on religious grounds, that is Jews or Muslims wishing to purchase kosher or halal meat respectively. Current marketing practice is for such meat to be marketed very largely through specialist outlets so that consumers wishing to buy this meat are able to do so. Whilst it appears that overall few consumers wish to have this information it may be that consumers are less well informed about slaughtering methods and the associated welfare issues, than they are about methods of rearing animals, which attract considerable publicity. In particular most consumers are unlikely to be aware that some meat from religiously slaughtered animals is diverted to the general market. We were aware that in contrast to the survey results, the response to the Government's consultation on the FAWC's 1985 Report on the Welfare of Livestock Slaughtered by Religious Method, indicated widespread and strong support among various interests, including some consumer groups, for a labelling requirement.

159. We looked specifically at whether there was a need to require labelling of religiously slaughtered animals since this was singled out by respondents to the consultation exercise and also because we were aware that the European Commission is working on a draft proposal for a Council Regulation on the protection of animals for slaughter which includes a requirement for all meat produced by Jewish and Muslim methods to be identified for the final consumer. However we believe that the same considerations and principles apply as those discussed above in relation to labelling to indicate methods of rearing. A labelling requirement would result in severe practical difficulties involving documentation and labelling right through the food chain. Whilst labelling might not be impossible for cuts of meat, meat products would be very difficult to control with considerable scope for evasion. Nonetheless we again believe that the major factor is that there is no physical difference between meat produced from religiously slaughtered animals and those slaughtered by conventional methods and that labelling is not the answer to welfare or ethical concerns, whether those concerns are justified or not. *We do not therefore recommend a requirement to indicate any slaughtering methods on food labels.* If it is necessary to address the concerns of consumers wishing to avoid buying meat from religiously slaughtered animals a better way of achieving this might be to constrain as far as possible the distribution of such meat to prevent it from reaching the general market.

SECTION VIII

TREATMENTS TO LIVING ANIMALS AND CROPS

BACKGROUND

160. The Committee was asked to consider whether treatments to living animals or growing or harvested crops with veterinary medicines or pesticides respectively should be indicated on the labelling of both prepacked and non-prepacked foods. As background to our consideration we looked into the current level of use of veterinary medicines and pesticides and at the various control measures in these areas.

161. We understand that as with humans most animals require some medication if they are to live healthily to and during maturity. During its lifetime the average animal may receive a variety of different medications, such as vaccinations against diseases and preventative and curative treatment against a variety of internal and external parasites as well as microbial infections. Another class of medicines, which we considered separately, are those, such as growth promoters which do not affect health, but are administered to animals with the deliberate intention of affecting physiological functions in a commercially desirable way.

162. In order to be offered for sale in the UK a veterinary medicine must be licensed under the Medicines Act 1968. The licensing procedure requires all scientific data about a product available to the company applying for a licence to be submitted to the Ministry of Agriculture, Fisheries and Food for independent assessment on the basis of safety, quality and efficacy. Safety includes safety to the treated animal, to persons handling the medicine, to the environment and, where the product is intended for administration to food-producing animals, to consumers of produce from that animal. Where the product is intended for use in food-producing animals a major part of the assessment focuses on consumer safety. Companies are required to conduct toxicological studies to demonstrate that an acceptable daily intake (ADI) may be set for each of the active ingredients of their products and at what level. From this a maximum residue limit (MRL) may be calculated. Companies are then required to demonstrate, by residue depletion studies, that food produced from treated animals will not contain residues in excess of the MRL. A withdrawal period may be determined in order to achieve this, again with a substantial precautionary margin. It is a legal requirement that the withdrawal period appears prominently on medicines labels along with any other information which users need to know to use the medicine safely. Where it is considered unsafe to use a particular medicine in food-producing animals, a contra-indication to this effect is included on labels. These measures are backed by the Ministry's programme of monitoring veterinary residues in animal products on farm and at slaughterhouses. Some 48,000 samples are taken each year and analysed for the presence of illegal substances or of permitted substances in excess of the MRL.

163. Pesticides are widely used on crops, for example to help improve yields, to protect crops and to extend the life of produce after harvest. The framework for control is broadly similar to that for veterinary medicines. Under the Food and Environment Protection Act 1985, Ministers must be satisfied that a product can be used safely before any approval for its sale, supply, use, storage or advertisement is given. In order to obtain approval pesticide manufacturers are required to provide a range of scientific data, which are scrutinised by the independent experts of the Advisory Committee on Pesticides (ACP). The Committee's examination includes assessment of studies on the short and long-term toxicity of the product, on its cumulative effects and on any delayed effects which might emerge after a latent period. Only when Ministers are convinced, on the basis of expert advice, that the product can be used without risk to people (whether operators, consumers or those in the vicinity of spraying operations), livestock or domestic animals and with minimal risk to the environment, will authority to put it on the market be given.

164. The approvals process is supplemented by statutory maximum residue limits (MRLs) laid down in the Pesticides (Maximum Residues in Food) Regulations 1988 which apply to both domestic and imported produce. MRLs are not safety limits but they provide an assurance that residues are within safe limits and that pesticides have been used correctly. MRLs are defined as the maximum amount of residue expected to arise if an approved pesticide is used according to the approved label instructions and they are set with the aim that likely intakes should be well within Acceptable Daily Intakes (ADIs). If the effective use of pesticide resulted in residue levels which were above safety limits, approval for its use would be refused or revoked, or its conditions of use amended. Monitoring is carried out by the Working Party on Pesticide Residues. Some 3000 samples are analysed each year for a wide variety of pesticides. The results of monitoring are published and show that pesticide residue levels in food in the UK are generally low and that estimated average dietary intakes are well within ADIs. Published results also provide consumers with information on the nature and level of residues which might be expected to arise in UK food supplies.

Legislative position

165. There is a requirement in the European Community Food Labelling Directive, as implemented in the Food Labelling Regulations 1984, for foods to carry a full list of ingredients. 'Ingredient' is clearly defined as meaning 'any substance, including additives, used in the manufacture or preparation of a food and still present in the finished product even if in altered form'. The legislation therefore draws a clear distinction between substances purposefully added to a food and those which come to be in the final food by other means, such as pesticide residues, and veterinary products used in animal husbandry. Only the first category is required to be identified on the food label. Regulation 12 of the Food Labelling Regulations requires any treatments or processes a food has undergone to be indicated in the name of the food if omission could mislead the consumer. This provision of the Regulations is usually taken to have a fairly restrictive interpretation, which is that the treatments covered are those which are analogous to the examples given in the Regulations and which change the nature of the food. There is nothing in the legislation to prevent voluntary indications concerning pesticide or veterinary medicine treatment. Such information would be subject to the general provisions of the Food Safety Act 1990 and the Trade Descriptions Act 1968, which prohibit false or misleading descriptions of food.

166. A European Community proposal on the labelling of organic foods and another on the labelling of post-harvest pesticide treatments have been issued. The proposed Council Regulation on the production and labelling of organic foods would make it illegal for a product to be labelled organic unless it had been produced in accordance with the standards set down in the regulation. The regulation is not now expected to be adopted until 1991. Initially it will not cover animal products. The draft European Community regulation on setting maximum pesticide levels for certain fruit and vegetables originally contained the proposal that post-harvest treatments should be labelled. We understand that the labelling requirement was removed from the text on the understanding that the whole issue of pesticide treatment labelling (both pre- and post-harvest, and for all produce not just fruit and vegetables) will be given separate and detailed consideration. The Council of Ministers is due to take a decision on this by 31 December 1991.

Views of interested parties

167. The trade registered strong opposition to any statutory requirement for the labelling of pre-or post-harvest treatment of crops. They pointed out that the labelling of specific pesticides would present considerable practical difficulties, as it would involve an elaborate system of recording information through the growing and harvesting stage to the retail end which, without considerable monitoring and checking, could not be guaranteed to give accurate and meaningful information to consumers. Any attempt to extend the requirements into the processed food market would, it was suggested, present insurmountable difficulties. Many trade respondents also pointed out that, as very little food was produced without the aid of pesticides, the requirements would apply almost across the board and therefore the only realistic approach (and one which would be more appropriate to consumer needs) would be to work within the context of regulated negative claims, which indicated that food had not been treated. Alternatively reliance could be placed on claims that foods were 'organically produced'. The trade also thought that labelling was unnecessary because of the strict controls on the approval and use of pesticides and over residues. Moreover they were concerned that labelling might be misleading and alarmist, and undermine consumer confidence in food particularly amongst those who were unaware of the controls. Some trade interests believed that any requirement could only be accommodated in the most general terms, for example 'treated with A or B' or 'may have been treated with C', or 'conforms to EC/UK legislation'. While this was put forward as a practical solution, doubts were also expressed about the value of such information to the consumer.

168. There was less comment from the trade on the question of labelling treatments to living animals with veterinary medicines, although essentially the same objections were made on the grounds that veterinary medicines were approved for use and subject to strict controls on use. The point was also made that veterinary treatments vary from animal to animal in the herd and throughout the life of the animal, and therefore it would be impossible to check over the medical history in order to satisfy the requirements. There was some acknowledgement that a labelling requirement in respect of non-therapeutic treatments, for example bovine somatotrophin (BST), might be a possibility but concern that such labelling, with details which would inevitably be of a technical nature, would be unnecessarily alarmist to consumers.

169. The majority of enforcement respondents doubted whether consumers wanted this information and suggested that most consumers were aware that pesticides and veterinary medicines were used on most farms. The potential difficulties of effective enforcement of a labelling requirement were highlighted. In common with the trade it was suggested that a simple system would be to label for the reverse situation and to ensure that adequate controls were in place for claims such as 'not treated with pesticide' or, alternatively, 'organic'. On this basis the consumer would know that products not able to make this claim had been produced according to current agrochemical practice. It was also suggested that there might be a place for labelling 'non-essential' treatments of crops and animals.

170. There was no clear view from consumer groups on this question, although a slight majority favoured some form of clear meaningful labelling, particularly of post-harvest treatments of foods. Such labelling it was suggested, should include clear instructions/warnings on washing or peeling the fruit or vegetable. A number of consumer groups were doubtful of the need for such labelling for the majority of consumers, and acknowledged the practical difficulties for the trade and enforcement authorities which would result from such a requirement. There was significant support for the view that the labelling emphasis should be placed on the untreated products. The proposed European Community controls on organic production and labelling were seen as a good example of how these types of claims could be controlled, allowing consumers to make proper choices in respect of treatment of crops and animals.

Consumer survey

171. Question 10 of the consumer survey aimed to establish which items, that could appear on labels in the future, respondents would really use and refer to. This was a prompted question which amongst other items (eleven in all) referred to whether pesticides are used; whether food is produced organically; treatment of live animals with veterinary medicines and the name of pesticide used. 50% of respondents said they would refer to and use information on whether pesticides had been used in fruit and vegetable production, placing this item second in the list. Only 23% expressed an interest in the name of any pesticides used in products, suggesting a general, rather than a specific, interest. Given an opportunity to have only one of the eleven items of information added to food labels, the item on whether pesticides had been used came third in the list but represented only 11% of all respondents. The name of the pesticide used came last with 3%. The survey shows that concern over health, or possible damage to health, was the most common reason (23%) why these 50% of respondents would refer to information on pesticides. Interest in what chemicals had been used; a general interest in pesticides; to know whether to wash the produce and identify naturally produced produce, were the other reasons given.

172. The treatment of live animals with veterinary medicines was selected by 31% of all respondents as information which they would refer to and use, placing it seventh in the prompted list of items. Concern about these medicines passing through the food chain and the possible associated health risks were the main reasons for selecting it. Next to this was the concern about the welfare of the treated animal.

173. 42% of all respondents selected the item about whether food is produced organically as information they would refer to and use. This places it higher on the list of items than either type of farming methods, veterinary treatments of animals, slaughtering methods or names of pesticides, but below information on whether pesticides are used. The healthier nature of the food, and concerns about too many chemicals in food were the main concerns of these respondents.

174. The survey results on these particular aspects of labelling should be considered against the fact that 75% of all respondents said that there was not any information which should appear on food labels that currently does not. It should also be noted that following on from this question no respondents (in an unprompted situation) mentioned information about pesticides or treatments of animals with veterinary medicines as information missing from current food labels.

Conclusions and recommendations

175. We recognised that the question of labelling foods to indicate treatments to animals or crops was different from the question of labelling to show rearing or slaughtering methods, discussed in the previous section, since 'permitted' residues may be present in the final food which is thus different from the untreated product. However we understand from recent published surveillance data that, when properly used, the majority of all pesticides used on crops do not leave residues in food and that the residues of many veterinary medicines will be completely broken down and, when the withdrawal period is observed, all residues depleted to very low concentrations before the food product is marketed. Moreover the treated product will be the 'norm' and across-the-board labelling, particularly of a general nature ('treated with pesticide') as requested by respondents to the consumer survey, would therefore not be helpful to consumers. Indeed it could be considered misleading and unnecessarily alarmist, especially when residues may not even be present. We felt that labelling of the untreated food or more detailed labelling (for example to indicate the type of medicine or pesticide used) would be more meaningful as it would enable consumers to distinguish between products and make positive choices.

176. We then looked carefully at the practical difficulties of meeting a more specific labelling requirement. As far as veterinary medicines are concerned, except in apiculture and the poultry industry, where all animals in a particular hive or poultryhouse tend to receive identical medication, the medical history of each animal is individual. Labelling on the use of veterinary medicines would therefore require the meat and milk of each animal to be identified, marketed and labelled separately. Even where animal produce is sold unprocessed this would require a complete and costly reordering of marketing arrangements; where the produce of several different animals is brought together and mixed, and particularly during food processing, it would be very difficult if not impossible to preserve separate identification. Food prices would be likely to rise.

177. Similar difficulties would occur in the case of labelling of pesticide treatments. Any labelling of produce would require retailers and all those further back in the food chain to be able separately to identify it from the moment it was treated. If pre-harvest treatments were included the task would be likely to prove impossible – for example in 1987, 99% of dessert apple trees in England and Wales were treated. On average 13 different active ingredients were applied pre-harvest (4 herbicides; 9 sprays to fruit and foliage including winter washes). On average one active ingredient was applied post-harvest but one crop surveyed received 3 different active ingredients.

178. A requirement to label post-harvest treatments only would be more manageable but would still require the introduction of complicated procedures to identify treatments and keep produce separate, would put up prices for all consumers and would be very difficult to enforce. In the UK post-harvest treatments are mainly restricted to potatoes, apples, pears, white cabbage, onions and carrots but there will be many imported food items which may have been treated post-harvest with a wide variety of pesticides. The costs to UK agriculture, packers, wholesalers, distributors and retailers would therefore be considerable even if labelling were restricted to post-harvest treatments. Increased costs from handling treated produce would arise mainly from the need to keep separate and identify produce from sources where different treatments have been applied. This would complicate the grading process and prevent the bulking of supplies. The increased costs directly associated with labelling would increase food prices.

179. Effective enforcement of any labelling requirement would also be difficult and costly but we believe it is not impossible if it is restricted to post-harvest treatments. Premium prices for food labelled as not treated with pesticides or veterinary medicines could well develop which would encourage unscrupulous producers and manufacturers to trade under fraudulent labels and steps would have to be taken to combat this. To prove a negative would be impossible, especially where residues are not normally present in the food, which is a reason for distinguishing between pre- and post-harvest treatments: pre-harvest treatments may or may not leave residues but post-harvest treatments are likely to be detectable.

180. We believe from the evidence of the consultation exercise and survey that it is a minority of consumers who positively wish to have information on veterinary medicines on food labels and that the practical problems and risk of consumers being misled rather than informed suggest that there should not be a labelling requirement. *We recommend accordingly.* However, we recognised that there might be a case for labelling non-therapeutic treatments because of their non-essential nature. In considering this separately we took account of the fact that in most cases the final food would be no different from that derived from animals which had not been treated for non-therapeutic reasons and the same practical difficulties outlined above would apply. Some consumers may be concerned and confused about the use of non-essential treatments, such as the use of growth promoters but we do not believe that labelling is the appropriate means to deal with these concerns. Indeed we believe it might well add to the concern and confusion. Such concerns should be addressed by ensuring that proper controls are in place and by consumer education about those controls and about the effect (or lack of any effect) on foods and consumers of the use of treatments.

181. On the question of whether foods from crops treated with pesticides should be labelled we took note of the message which emerged from both the consultation exercise and consumer survey that a significant proportion of consumers would welcome some general information on the use of pesticides in food production, although there appeared to be considerably less interest in the specific pesticides used. The consultation exercise also indicated that there may be greater concern on the part of consumers in respect of post rather than pre-harvest treatment of fruit and vegetables, as consumers wish to be able to handle and prepare the treated food appropriately. we also noted the view that labelling for the absence of treatments either by veterinary medicines or pesticides presented an alternative and practicable approach. The sizeable proportion of consumers in the food labelling survey who mentioned organic production as information they would refer to and use on food labels, might be seen as supporting this approach.

182. We considered pesticide and post-harvest treatments in the widest sense of those terms and looked into the possibility of a number of different options:

(i) no labelling at all, on the assumption that those who do not want produce treated with synthetic pesticides increasingly have the option of buying organic produce;

(ii) labelling of post-harvest treatments only, as proposed by the European Commission, on the basis that these treatments are most likely to leave residues in food;

(iii) labelling of pre- and post-harvest treatments where residues are left;

(iv) labelling of all pesticide treatments whether or not residues are left.

183. In addition to the practical difficulties and costs involved, which might work to the detriment of the majority of consumers, we were concerned that labelling might be potentially confusing for consumers. We believe that the labelling of all treatments would be meaningless if the requirement was for a general statement only, since it would apply to most foods. Whilst more specific details of the pesticide used might be considered informative, many consumers may be confused by complicated scientific names for active ingredients. (We understand that in the UK alone over 400 active ingredients are approved for agricultural use). A 'P' number system (as has been suggested) could equally be confusing. Consumers may conclude that something new and unsafe has been added to their food and, as with 'E' numbers, some consumers may simply avoid foods where a treatment with a pesticide is indicated. This may not be in their best interests as far as achieving a healthy, balanced diet is concerned.

184. However, we feel it is important that all substances deliberately added to the final food should be declared to the consumer. This is the approach we adopted in relation to ingredient listing (Section IV). We believe that there is merit in distinguishing between treatments such as waxes on fruit which are purposefully added post-harvest to protect the properties of the final food as sold to the consumer and those pesticides, for example use of a fungicide on the growing crop to prevent mould, which may remain adventitiously in a food from a pre-harvest application. We recognise that labelling of post-harvest treatments only might mislead consumers into thinking that produce which was not labelled had not been treated with pesticides when in fact it might have been treated pre-harvest, possibly even with the same active ingredient as post-harvest treatments, and might contain residues. *Nevertheless despite the difficulties we believe that there is a need to indicate all post-harvest treatments where the produce itself is the food rather than an ingredient. Labelling or appropriate notices in the case of non-prepacked foods should make clear that post-harvest treatment has been used and the labelling should be along the lines of 'post-harvest pesticide/treatment used'. Educational campaigns should also be introduced to help prevent consumer confusion.*

185. We believe that the positive and controlled labelling of organic produce will also improve consumer choice in this area . We welcome the progress that has been made so far through the standards developed by the United Kingdom Register of Organic Food Standards and on the European Community Regulation on the production and labelling of organic foods which, however, does not cover products of animal origin.

186. *Whilst we recognise that when a pesticide is approved, safety calculations are made on the basis that consumers will eat treated produce unwashed and unpeeled, we nevertheless understand consumer concern about being able to prepare appropriately produce which has been treated post-harvest, for example by peeling or washing thoroughly, and recommend that this information should also be given to consumers.* It is already a requirement of the Food Labelling Regulations that any special conditions of use or instructions for use should be given for prepacked foods. For non-prepacked foods, such as fruit and vegetables, in Section IX of this Report we encourage the industry to give voluntarily as much information as possible, in particular information such as instructions for storage and use.

187. Consumer concern over safety and health appeared in the survey as the most common reason for wanting labelling of pesticide treatments. As already stated our view is that the purpose of a labelling requirement is primarily to give information which the consumer needs to know about the final food, to distinguish it from other products with which it might be confused and to prevent the consumer being misled; hence our recommendation for the labelling of post-harvest treatments only. Labelling should not be used primarily to allay concerns or to educate. *We believe that public concern should be addressed through a high profile, widespread education campaign, including action in schools and at European Community level.* As with veterinary medicines, consumers should be encouraged to understand that, unless otherwise stated (for example in the case of organic agriculture), the food would have been produced using pesticides in a conventional agricultural system and that such pesticides should have been used safely in accordance with the various legislative controls or Codes of Practice backed up by food surveillance. Where post-harvest treatments are not indicated consumers should be educated to understand that this does not necessarily mean that pesticides have not been used pre-harvest nor that the food is residue-free.

SECTION IX

FOODS PRODUCED USING GENETIC MODIFICATION

188. The labelling of foods produced using genetic modification did not form part of our review of food labelling legislation and practices since consideration of this very important issue was already underway before this review began. Nevertheless we believe that it is useful to include our present views in this report.

189. Ministers have specifically asked us to consider, on a case by case basis and in the context of the Food Labelling Regulations 1984, the need for special labelling of foods produced using genetic modification following their evaluation by the Advisory Committee on Novel Foods and Processes. As part of our general consideration of this issue we took into account those uses of this technology which could have moral and ethical implications for some consumers. We felt that labelling was not the answer to these concerns except where the presence or use of genetically modified organisms could be considered to alter materially the nature of the food. We concluded that under these circumstances specific food labelling – 'contains the products of gene technology' – should be used to inform consumers of the use of gene technology. To assist us in the specific task of deciding for each new product whether special labelling is required we have drawn up criteria which take particular account of whether the final food is different from that produced by conventional methods. These criteria presently consist of four basic food categories which act as a primary screening mechanism to indicate whether such labelling is necessary. These categories have formed the basis of the Committee's Guidelines for the Labelling of Food Produced Using Genetic Modification (Appendix VI).

190. In brief summary, our advice is that consumers would probably need to be informed by labelling in all cases where the food contains either a trans-species genetically modified organism (that is, formed from the genetic material of different species) or its genetic material. In addition we advised that novel food products of a genetically modified organism would also require specific labelling of the food. For other situations, such as foods which contain neither the genetically modified organism, nor its DNA, or foods which contain an intra-species genetically modified organism (that is, formed from the genetic material of the same species) then no such labelling would be required. We stress, however, that each case will need to be considered on its merits and that as a consequence the guidelines may then need to be amended.

SECTION X

NON-PREPACKED FOODS INCLUDING FOODS SOLD BY CATERERS

BACKGROUND

191. We considered what essential information needs to be given with foodstuffs which are sold loose. This includes non-prepacked food sold at retail level, including foods prepacked by retailers for direct sale on the premises and foods for immediate consumption sold at catering establishments.

Legislative position

192. Both European Community and UK food labelling legislation apply to all foods, non-prepacked as well as prepacked, but within the framework non-prepacked foods are exempt from most requirements. Under the European Community Food Labelling Directive the detailed rules which should apply to non-prepacked retail sales and to foods sold in the catering sector are at the discretion of Member States, but with the important proviso that the consumer still receives sufficient information. In the UK, in recognition of the practical difficulties in labelling, or otherwise giving information on food sold in this way, a minimum of information is required to be given.

193. Foods sold non-prepacked, such as delicatessen foods, foods sold loose on market stalls, or which are prepacked by a retailer for direct sale, for example, bread baked on the premises and wrapped, generally need only be marked or labelled with the name of the food and the category name of the following types of additives present in the food: antioxidant, artificial sweetener, colour, flavour enhancer, flavouring, preservative. These particular categories of additives were selected following a recommendation by the Food Standards Committee in its Second Report on Food Labelling (1979) as those concerned with ensuring that food is presented in a sound and attractive condition, but which might give a misleading impression to the consumer as to the nature or quality of the food. Edible ices, flour confectionery or bread prepacked by a retailer for direct sale by him, or flour confectionery which is packed in a crimp case or unmarked wholly transparent containers need not be marked with any labelling particulars except that edible ices and bread (except white bread) must be marked, or labelled with the name of the food and all must bear an appropriate indication if they contain any of the categories of additives listed above.

194. The required labelling particulars may be given on the label attached to the food or on a ticket or notice displayed in immediate proximity to the food, when sold to the ultimate consumer, and may be given in relevant trade documents on or before delivery when sold to a caterer.

195. Non-prepacked foods sold at catering premises for immediate consumption there are not required to be labelled with any particulars. However, under general controls in the Food Safety Act 1990 and Trade Descriptions Act 1968 any descriptions which are applied must be accurate and not misleading. 'Catering establishment' is widely defined in the legislation and includes, for example, restaurants, schools, hospitals, clubs or any establishment where in the course of a business food is prepared for the consumer. Small prepacked individual portions, such as jam or butter intended as an accompaniment to another food or any prepacked sandwich, filled roll or similar bread product or prepacked, prepared meal, must give the name of the food. Non-prepacked and similar foods sold for immediate consumption, other than at the place where sold, for example 'take away' foods, need only give the name of the food. Special manner of marking requirements very similar to those for non-prepacked retail foods apply.

Views of interested parties

196. Trade interests generally felt that the disparity within the current labelling requirement for prepacked and loose retail food sales was disadvantageous to the consumer in preventing proper comparisons to be made. In many cases additional labelling requirements were sought up to the same level as prepacked foods but with some acknowledgement given to the particular practical difficulties involved. The trade were clearly of the view however that the 'practical difficulties' argument was less defensible today, given current information handling technology. Food prepacked for direct sale, in particular in-store bakery bread, which often carried voluntary labelling or promotional information was mentioned as an area of particular concern, and one which exemplified how the present requirements had fallen behind developments, leading to unfair competition for the trade and insufficient information for consumers.

197. Enforcement and consumer interests were similarly of the view that the consumer was entitled to the same labelling information for non-prepacked retail sales, including 'take away' foods, as for prepacked foods. The point was made that loose food marketing and 'pack your own' sales have grown considerably since the regulations were framed and were still increasing. In addition, it was felt that the general move towards more informative labelling (rather than strict compositional controls) must be reflected in the non-prepacked foods area. Enforcement interests were fairly confident that the practical difficulties of presenting the information could be overcome. Special mention was made of the declaration of additives in foods sold this way and the need for full declarations, including E numbers and specific names, rather than the present selective approach. Foods sold in catering establishments for immediate consumption were considered to present the greatest practical difficulties in providing detailed information.

Consumer survey

198. The food labelling survey closely reflected the views of respondents to the consultation exercise on the types of additional information, which it was felt should be required for loose retail food sales. Over two thirds of survey respondents indicated that they would like additional information with food sold loose. A date-mark was top of the list of mentioned items (38%) with ingredient listing, place of origin, nutrition information and storage instructions next in order of priority. For catering sales about half of those surveyed (49%) said that no additional labelling information was necessary. Of those requesting some additional labelling a list of ingredients and the cooking method came top of the list of requests but these were mentioned by only 16% and 14% of respondents respectively. The menu was by far considered to be the most suitable place for giving the additional information to consumers.

Options considered and recommendations

199. We recognise and have sympathy with the clear concern expressed by the majority of respondents that in principle, the present labelling requirements for loose foods sold at retail level are unsatisfactory, as they do not represent the essential information which most consumers wish to receive about food sold this way. We also share the view of many respondents that the needs of the consumer for information on loose foods are much the same as for prepacked foods and that there is a need to address the present disparity in the amount of information provided to consumers in these two areas. There is good reason in principle why the information required to be given on prepacked foods should be extended to cover loose foods. Whilst we hope to see considerable progress in this area, with every effort being made by the industry to supply more information, we recognise the need to take account of the practical difficulties with which some sectors would be faced if wide legislative requirements were introduced.

200. We looked very carefully at the practical implications for all sectors of the trade of requiring labelling at the same level as for prepacked foods. Many respondents, in recognition of the practical difficulties involved, suggested that consumer needs could be met by an intermediate level of labelling to give certain key particulars such as date-mark, storage instructions, conditions or instructions for use, according to the particular type or category of food on offer. We believe that this is a sensible approach. A requirement for labelling at an intermediate level, which would also allow the trade a degree of flexibility in choosing which information is most essential and appropriate for the particular foods on offer and the means to convey that information, would be difficult to achieve through statutory means given the wide variety of foods available. *We decided therefore against recommending further statutory controls in this area.* We would however *strongly encourage the trade to respond to consumers' wishes for more labelling information for non-prepacked foods and suggest that at least the key particulars mentioned above, depending on the type of food, should be supplied.* We believe that industry should be able to adapt present labelling procedures in these areas to provide clear, essential information to the consumer, for example by giving key safety information on a poster ('our meat products should be kept refrigerated below 5°C and used within 2 days of purchase') and ingredient lists could be provided in a leaflet on the counter.

201. On one particular aspect, namely the present requirements of the legislation for certain categories of additives to be indicated for non-prepacked foods, we agree with respondents that there is little justification for continuing with this selective approach. The consultation exercise and consumer survey have both shown that consumers wish to have full and clear information about additives in food and therefore find it less than helpful to have only certain categories listed and no indication of the particular additive used. It is likely that many consumers believe the list of additive categories indicated for these foods to be a comprehensive one and are therefore misled or confused by partial information. We also took account of the fact that improvements in this area could be brought about by simply extending the present requirements of the regulations without significant practical difficulties for the trade. *We therefore recommend that the present requirements of the Regulations, for certain categories of additives to be indicated should be extended to require all categories to be listed.* Where there is no appropriate category name specified in the Regulations *we recommend, in line with the recommendation we have made in respect of the labelling of additives in food, that a meaningful descriptive function name should be given. We further recommend that where the retailer, in response to the Committee's recommendation for fuller labelling of non-prepacked foods, voluntarily provides more detailed information, including a list of ingredients in the food, then any additives present should be indicated in line with the requirements of the regulations in respect of listing additives in prepacked foods.* This means that the function name together with the E number or specific name or both should be given in line with our recommendations in paragraph 96.

202. The Committee looked carefully at the concept of *foods prepacked for direct sale*, such as in-store bakery products, where respondents indicated that the present requirements for the labelling of such foods had fallen out of step with developments in foods sold this way. We recognise that the exemption was originally adopted as a sensible means of ensuring that small 'corner shops', producing a limited range of their own products on the premises, were not unduly burdened by detailed labelling requirements which might seriously affect their viability. We consider however that the decline in the number of such operations, together with the significant increase in 'prepacked for direct sale' foods offered by the larger retailers, has resulted in an anomalous situation as regards the amount of labelling information provided to consumers purchasing very similar foods often in close proximity to each other. We have taken account of the fact that many such large retailers voluntarily provide additional labelling information, including promotional material, on these foods, which suggests that the practical difficulties of providing the full range of information are not great. *We therefore recommend that the requirements for the labelling of foods prepacked for direct sale should be extended to the same level as for prepacked foods.* The Committee nevertheless recognises that the variation in products sold in this way may give rise to certain practical difficulties, particularly as regards providing a detailed list of ingredients in ingoing order of weight, and that these will be considerably more significant for the smaller retailer. *We therefore recommend that the extended labelling requirements should be framed so as to allow 'typical value' indications of the ingredients for such products.*

203. Finally, we considered *foods sold by caterers* where we recognise that respondents did not indicate any strong desire among consumers for more detailed labelling of such foods. we also recognise that there are considerable practical difficulties for labelling foods sold by caterers particularly where the food is in the form of a meal prepared on the premises. For prepacked foods sold by caterers, such as sandwiches, filled rolls and other similar products, where the regulations already require full labelling when these products are supplied to the caterer (although this may be given in associated trade documents) *we consider that caterers should pass all the information on to the consumer. For other catering foods we do not consider that changes to the legislation are needed although we would strongly urge caterers to provide the fullest possible descriptions of the food on offer to the consumer.*

APPENDIX 1

LIST OF ORGANISATIONS WHICH RESPONDED TO THE CONSULTATION EXERCISE

FOOD ADVISORY COMMITTEE REVIEW OF FOOD LABELLING

Agriculture and Food Research Council Institute of Food Research
Association of Public Analysts
Association of Public Analysts of Scotland
Association of Scottish Shellfish Growers
Biscuit, Cake, Chocolate and Confectionery Alliance
Boots Company Limited
British Agrochemicals Association Ltd
British Dental Association
British Meat Manufacturers Association
British Nutrition Foundation
British Paedodontic Society
British Soft Drinks Association Ltd
British Sugar
British Veterinary Association
Campaign for Real Ale Ltd
Campaign for the Protection of Shechita
Co-operative Union Ltd
Coeliac Society of the UK
Consumers' Association
Consumers in the European Community Group (UK)
Convention of Scottish Local Authorities
Cooks Town
Coronary Prevention Group
Cow and Gate Ltd
Doves Farm
Dudley Metropolitan Borough
European Communities Economic and Social Committee
Farm Animal Welfare Committee
Federation of Bakers
Fish and Meat Spreadable Products Association
Food Additives Industry Association
Food and Drink Federation
Food and Health Policy Research
Fresh Fruit and Vegetable Information Bureau
Gin Rectifiers and Distillers Association
Health Promotion Authority for Wales
Ice Cream Alliance
Incorporated National Association of British and Irish Millers Ltd
Independent Television Association
Institute of Food Science and Technology (UK)
Institute of Trading Standards Administration
Institution of Environmental Health Officers
International Association of Fish Meal Manufacturers
Kellogg Company of Great Britain Limited
Leatherhead Food Research Association
Local Authorities Co-ordinating Body on Trading Standards
London Food Commission

Margarine and Shortening Manufacturers Association
McDougalls Catering Foods Ltd
Meat and Livestock Commission
Midland Trading Standards Liaison Group
Milk Marketing Board
National Association of Citizens Advice Bureaux
National Association of Master Bakers, Confectioners and Caterers
National Consumer Council
National Council of Women of Great Britain
National Farmers' Union
National Farmers' Union of Scotland
National Federation for the Blind
National Federation of Consumer Groups
National Federation of Fruit and Potato Traders Ltd
National Federation of Meat Traders
National Forum for Coronary Heart Disease Prevention
National Housewives Association Ltd
National Office of Animal Health Ltd
Newry and Mourne District Council
Northern Ireland Coronary Prevention Group
Potato Marketing Board
Produce Packaging and Marketing Association Ltd
Retail Consortium
Robert Gordon's Institute of Technology
Royal Environmental Health Institute of Scotland
Royal Society for the Prevention of Cruelty to Animals
Scotch Whisky Association
Scottish Association of Master Bakers
Scottish Federation of Meat Traders Association
Scottish Grocers Federation
Shropshire County Council
The Advertising Association
The Board of Deputies of British Jews
The Brewers Society
The British Dietetic Association
The Vegan Society
The Vegetarian Society of the UK
Ulster Farmers Union
Union of Orthodox Hebrew Congregations
United Kingdom Agricultural Supply Trade Association
United Kingdom Federation of Business and Professional Women
United Kingdom Home Economics Federation
United Kingdom Provision Trade Federation
University of Newcastle upon Tyne
Waitrose
West Midlands Joint Committee – Trading Standards Sub-Committee
Western Health and Social Services Board
Womens Farmers Union
Womens National Commission
World Society for the Protection of Animals

APPENDIX II

SUMMARY OF FOOD LABELLING CONSUMER SURVEY RESULTS

EXTRACT

FOOD LABELLING SURVEY ENGLAND AND WALES
Ministry of Agriculture, Fisheries and Food
(HMSO: ISBN 0 11 242904 1)

BACKGROUND

A number of recent issues have resulted in pressure for changes in the way food products are labelled in the UK. For example, an upsurge in awareness of and interest in the environment has led to demands for proposals for more detailed information on food in the area of 'green' or ethical issues, such as production methods. Concerns about food safety have highlighted the public's awareness of food labelling, particularly date-marking, storage and cooking instructions. The trend towards healthier eating has also increased consumer demand for more detailed, accurate and accessible information, in particular on nutritional content, ingredient listing and claims.

The Ministry of Agriculture, Fisheries and Food (MAFF) has produced a booklet for consumers on understanding food labels, but believes that certain aspects of food labelling remain confusing for many consumers. The Food Minister has therefore asked the independent Food Advisory Committee to review food labelling legislation and practices to see how they can be best developed to give consumers what they need and want to know. Food labelling regulations in the UK are largely based on European Community legislation.

RESEARCH METHOD

1. The sample

A systematic random sample of 67 parliamentary constituencies in England and Wales, stratified in terms of population density, was drawn to act as interviewing points for the survey. The constituencies used, which were selected to provide a sample matching the regional profile of England and Wales, are listed in Appendix A. Fifteen interviews were required in each constituency to give a final sample of at least 1000.

Previous research has shown that certain groups show more interest in food labelling and food in general than do others. Examples of such groups are: mothers of small children; the better off financially; certain age groups; and some religious and ethnic minority groups which prohibit the consumption of certain foods. To ensure that the views of such individuals would be obtained, quotas were set on sex, age, social class and the presence of children in the household. In addition, interviewers were required to interview three members of religious or ethnic minorities in each constituency.

To qualify for the survey, in addition to meeting the quota requirements, respondents (all of whom are aged 18 or over) had to be solely or mainly responsible for shopping for food and drink for their household, or to share this responsibility with someone else. For this reason the sex quota was set to produce a sample of approximately three quarters women, one quarter men.

2.   The questionnaire and pilot survey

From a question list prepared by COI a pilot questionnaire was drawn up by PAS, and the content and contacting procedure tested in a small scale pilot exercise. One interviewer worked one assignment (15 interviews) in Ilford in the week preceding 19 April, and on this date was personally debriefed on her experiences. The questionnaire was quite substantially modified in light of the pilot survey results, and a copy of the final version of this document (which took approximately 30 minutes to administer) can be found in Appendix B.

3.   Main fieldwork

Interviewing was carried out in the constituencies listed in Appendix A between 25 April and 6 May 1990.   Full written instructions on the contacting procedure and the questionnaire were sent to all interviewers working on the survey with their fieldwork materials.

In addition to the questionnaires and showcards, interviewers were provided with copies of a letter of authorisation from COI, to be shown to respondents when initial contact was made. This letter simply confirmed that PAS were carrying out the project for COI, assured respondents of the confidentiality of their responses and gave contact names and telephone numbers in the event of any queries about the survey. A copy of the letter can be found in Appendix C.

The total number of interviews achieved was 1028.

Summary of main findings

● Two thirds of respondents claimed to be solely or mainly responsible for their household's food and drink shopping: more women than men (77% versus 36%). Older respondents and those from social grade DE were more likely to claim responsibility.

● Just under a quarter (23%) shopped at least four times a week, and the vast majority of respondents (96%) shopped once a week or more often.

● Hypermarkets or large supermarkets were the most frequently mentioned type of outlet used for the bulk of shopping: 26% used such a store outside town centres most often, and 36% used the same type of outlet in a town centre. Taking into account all the outlets used for food and drink shopping, over half (52%) of the sample used a small local or corner shop regularly.

● Specific religious controls or restrictions had to be considered when food shopping by 12% of respondents, vegetarian or vegan diets by 11%, and slimming diets by 11%. Avoiding allergy, high blood pressure or other medical reasons for dietary restriction were mentioned by 5% or less. Over half of the sample (57%) did not have to take any such restrictions into consideration when shopping.

● More women and those in social grades AB and C1 claimed to take notice of what is printed on food and drink labels than did men and the lower social grades. Overall 46% said they took a great deal or quite a lot of notice of such information.

- When shopping or in the home, 38% of respondents said, without prompting, that they particularly looked out for a list of ingredients on a food label, while 23% and 22% respectively looked for a datemark and nutrition information. Price and brand or manufacturer were looked for by 13% and 10%. When shown a prompt list of such information four fifths (79%) said they looked out for the datemark. Women tended to look for information on what the product contained, nutritional values and freshness, while men were more interested in price.

- Two thirds (67%) found none of the information that appears on food labels confusing. The list of ingredients was a problem for 11% however, and the nutrition information for 8%.

- Specific points of confusion mentioned were E numbers, complex descriptions or chemical names, and the relationship between imperial and metric weights and volumes.

- Asked which item of information on food labels was most important, a third (36%) chose the datemark, 18% the list of ingredients, 12% the price and 11% nutrition information. Two thirds put the datemark in their 'top three' of important items.

- A third (34%) agreed that they were not sure of the difference between 'best before' and 'use by' dates (more men than women), while 72% agreed that nutrition information made it easier to choose food for a healthy, balanced diet, in this case, more women (77%) than men (58%).

- Almost two thirds (64%) disagreed with the statement that there was too much information on food labels, while four fifths (80%) agreed that they would like the quantities of main ingredients shown. Over half (56%) agreed that some health claims made for products were difficult to believe. Half agreed that as they know which products they liked, the quantities of ingredients were not of interest to them.

- A third (36%) thought the information on labels was not clearly presented, and of these, half (52%) complained about small print, this being the most common complaint.

- Three quarters felt that no information was currently missing from food labels. Of those who would like more information, 18% would have liked to see a full list of ingredients, and 13% more information on E numbers.

- Respondents chose from a list of items of information those they would use: 52% would refer to quantities of main ingredients; 50% whether pesticides had been used in the production of the food; 45% more detail in general; and 42% whether the food had been produced organically. More women chose each item than men.

- Turning to food sold loose in supermarkets and delicatessens, 38% felt that a datemark should appear, and 16% a list of ingredients. A tenth would like the place of origin displayed, and a further tenth some nutrition information.

- Half felt that no further information on food sold in cafes or restaurants should be shown. Among those who would like to see more detail, the ingredients and cooking method were most frequently chosen. Of those who would like such information most (66%) thought it should appear on the menus.

● Over two thirds of respondents thought more general publicity about food labelling and what the items that appear on labels mean was desirable, and the most frequently mentioned source for such publicity being leaflets available through supermarkets. Television programmes and advertising were thought the best media by 36% and 29% respectively, with a further fifth suggesting newspapers or leaflets through the door.

Note: Appendices and social groups referred to in this extract are appendices and social grades as defined in the full report.

APPENDIX III

CURRENT AND FORTHCOMING FOOD LABELLING LEGISLATION, CODES OF PRACTICE AND GUIDELINES

A. GENERAL FOOD LABELLING

1. Food labelling is controlled in England and Wales by the Food Labelling Regulations 1984 (SI No 1305) (FLR), as amended by the Food Labelling (Amendment) Regulations 1989 (SI No 768). Parallel legislation applies in Scotland and Northern Ireland. The regulations implement respectively, Council Directives 79/112/EEC (food labelling), 77/94/EEC (PARNUTS) and 86/197/EEC (alcoholic strength labelling). The regulations apply to food which is ready for delivery to the ultimate consumer or to catering establishments.

Prepacked foods

2. When sold prepacked the food must be labelled with the following:

   (a) the name of the food;

   (b) a list of ingredients;

   (c) an indication of minimum durability;

   (d) any special storage conditions or conditions of use;

   (e) the name and address of the manufacturer, or packer, or EC seller;

   (f) place of origin, if omission would materially mislead with regard to its origin;

   (g) instructions for use, if appropriate use could not be made of the product without those instructions;

   (h) for drinks with an alcoholic strength of more than 1.2% alcohol by volume an indication of alcoholic strength by volume (in the form of a figure and the symbol '% vol.').

3. **The name of the food**
(FLR Regs 9-12, EC DIR. 79/112 Art.5)

   (a) If there is a name laid down in regulations, for example 'wholemeal bread' or 'marmalade', this must be used, if not a customary name may be used. In other cases a name precise enough to indicate the true nature of the product and distinguish it from others must be used.

   (b) The name of the food must include or be accompanied by an indication of its physical condition (e.g. powdered) or treatment (e.g. dried, frozen or concentrated) where a purchaser could be misled by the omission of that information. Special rules apply to frozen meat and offal, tenderised meat, processed peas and food frozen with dichlorodifluoromethene.

   (c) Schedule 1 to the Regulations contains special rules concerning names used for fish, melons and potatoes and vitamins.

4. **List of ingredients**
   (FLR Regs 13-20, EC DIR. 79/112 Art. 6 and ANNEXES 1 and 2)

   (a) A complete list of ingredients generally in descending order of ingoing weight is required. The name used for an ingredient must generally be a name which could be used if the ingredient were being sold as a food by itself. Added water must be shown in most cases when it exceeds 5% of the finished product.

   (b) Schedule 3 to the Regulations contains a list of generic terms such as 'fat', 'oil', 'meat', 'herbs' which may be used in an ingredient list, subject to certain conditions.

   (c) There are special rules for compound ingredients, for example, the names of the ingredients of a compound ingredient need not be given where the compound ingredient constitutes less than 25% of the finished product. However any additive which is present as an ingredient must be named.

   (d) Additives used as ingredients (except flavourings) must be described by the appropriate category name of the function performed in the food, for example, colours, stabilisers, preservatives, followed by their specific name or serial number or both. Where no category name is available for the function performed by an additive in a food, the additive must be declared in the ingredients list by its specific name.

   (e) Ingredients which need not be named include constituents of an ingredient temporarily separated in the making of the food and re-introduced in the original proportions; additives contained in an ingredient and serving no significant technological function in the finished product; additives used solely as processing aids; and any substances other than water used in an essential quantity as a solvent or carrier for an additive.

   (f) Foods which need not bear a list of ingredients include:

   - unprocessed fresh fruit and vegetables
   - carbonated water
   - vinegar
   - cheese, butter, fermented milk and fermented cream
   - flavourings
   - single ingredient foods
   - alcoholic drinks above 1.2% abv.

   (g) Where the presence or low content of an ingredient is given special emphasis an indication of the minimum or maximum percentage, as appropriate, of that ingredient must be given.

5. **Indication of minimum durability**
   (FLR Regs 21 and 22, EC DIR. 79/112 Art. 9)

   (a) The main form of indicating the minimum durability of the food, is 'best before' with the date until which the food can reasonably be expected to retain its specific properties if properly stored. For perishable foods intended to be consumed within 6 weeks of being packed the minimum durability may be indicated by 'sell by' immediately preceded or followed by the period from date of purchase for which the food can reasonably be expected to retain its specific properties if properly stored. Whatever form of datemark is used, any storage conditions which need to be observed if the food is to keep for the expected period must also be given.

(b) Some prepacked foods are not required to bear an indication of minimum durability, these include:

- fresh fruit and vegetables, including potatoes, which have not been peeled or cut into pieces;

- cider, perry, liqueur wine, sparkling wine, and aromatised wine;

- any alcoholic drink with a strength by volume of 10% or more;

- beer sold in a bulk container for resale;

- flour confectionery and bread intended to be consumed within 24 hours of preparation;

- deep frozen (including quick frozen) foods;

- foods with a minimum durability of more than 18 months;

- cheeses which ripen in the packaging.

6. Exceptions to the general rules for prepacked foods
(FLR Regs 25, 30 and 31, EC DIR. 79/112 Art. 13)

(a) Less onerous labelling requirements apply to individually wrapped fancy confectionery products and foods in small packages the largest surface of which has an area of less than 10 square centimetres.

(b) There are additional rules for food sold from vending machines, and for alcoholic drinks.

Non-prepacked and similar foods
(FLR Regs 23 and 24, EC DIR. 79/112 Art. 12)

7. Foods which are not prepacked, or prepacked by a retailer for direct sale need only be marked or labelled with the name of the food and the category name of the following additives if present in the food: antioxidant, artificial sweetener, colour, flavour enhancer, flavouring, preservative.

8. Edible ices, flour confectionery or bread, prepacked by a retailer for direct sale by him or flour confectionery which is packed in a crimp case or unmarked wholly transparent container, need not be marked with any labelling particulars, except that edible ices and bread (except white bread) must be marked or labelled with the name of the food and all must bear an appropriate indication if they contain any of the categories of additives in paragraph 7 above.

9. The required labelling particulars may be given on a label attached to the food or on a ticket or notice displayed in immediate proximity to the food, when sold to the ultimate consumer, and may be given in relevant trade documents on or before delivery, when sold to a caterer.

Food for immediate consumption
(FLR Regs 28 and 29, EC DIR. 79/112 Art. 12)

10. Non-prepacked foods sold at catering premises for immediate consumption there are not required to be labelled with any particulars. However, under the general principles of food and trade descriptions law any descriptions which are applied must be accurate and not misleading. Small prepacked individual portions, such as jam or butter, intended as an accompaniment to another food or any prepacked sandwich, filled roll or similar bread product, or prepacked prepared meal must give the name of the food.

11. Non-prepacked and similar foods sold for immediate consumption, other than at the place where sold, need only give the name of the food. Special manner of marking requirements similar to those in paragraph 9 above, but with some variations, apply.

Manner of marking requirements
(FLR Regs 32-35, EC DIR. 79/112 Art. 11)

12. In general for prepacked foods the labelling particulars must be shown on the packaging, or on a label attached to the packaging, or on a label clearly visible through the packaging when sold to the ultimate consumer; particulars may be provided in relevant trade documents on or before delivery when the food is sold to a catering establishment. In all cases the particulars must be easy to understand, clearly legible and indelible, and when the food is sold to the ultimate consumer, they must be in a conspicuous place so as to be easily visible. They must not be hidden, obscured or interrupted by any other written or pictorial matter. Where a datemark and/or the net quantity is required, they must appear in the same field of vision as the name of the food.

SPECIAL REQUIREMENTS FOR ALL FOODS

13. Misleading descriptions
(FLR Regs 38, 39 and Schedule 7, EC DIR. 79/112 Art. 2)

There are special conditions to be observed when certain words and descriptions are used in the labelling or advertising of foods. Some of the terms to which such conditions apply are: butter, cream, dietary or dietetic, flavour/flavoured, chocolate, fresh, garden, green, milk, alcohol-free, dealcoholised, non-alcoholic, shandy, shandygaff, ginger beer shandy, cider shandy, cider shandygaff, cider and ginger beer shandy, sweetened liqueur, vintage, Scotch whisky, Irish whiskey, blended Scotch whisky and blended Irish whiskey.

14. Claims
(FLR Regs 36, 37 and Schedule 6 EC DIR. 79/112 Art. 2)

Foods which claim to provide energy, contain vitamins or minerals, polyunsaturated fatty acids or cholesterol, aid slimming or weight control, or be suitable for diabetics, must be labelled with additional particulars that support the claim.

15. Foods for a particular nutritional use

Claims that a food fulfils a particular nutritional use cannot be made unless the food does in fact fulfil the claim and the food is labelled with an indication of the particular aspect of its composition or manufacturing process which gives it its particular characteristics. Furthermore if the food has been specially made it must also be labelled with the prescribed energy statement and sold to the ultimate consumer prepacked.

16. GENERAL REQUIREMENTS
Food Safety Act 1990 (Section 15); Food Labelling Regulations (Reg. 4)

The labelling, advertising and presentation of foods must not be such that a purchaser is likely to be misled to a material degree as to the nature, substance or quality of the food.

SPECIAL REGULATIONS

17. Certain foods are subject to specific regulations, national and EC in origin, which deal with the composition and labelling of those foods. The regulations fall broadly into two groups. The first includes those covering foodstuffs which are specifically excluded, for labelling purposes, from the Food Labelling Regulations 1984. (The following is not an exhaustive list.) These include:

- The Specified Sugar Products Regulations 1976 (implements 73/437/EEC)

- The Cocoa and Chocolate Products Regulations 1976 (implements 73/241/EEC)

- The Honey Regulations 1976 (implements 74/409/EEC)

- The Condensed Milk and Dried Milk Regulations 1977 (implements 76/118/EEC)

- The Coffee and Coffee Products Regulations 1978 (implements 77/436/EEC)

- Hen eggs, covered by European Community Marketing Regulations (2272/75/EEC)

- Wines, covered by European Community Regulations on the description and presentation of wines (355/79/EEC)

- Additives sold as such covered by Additives Regulations

With regard to the labelling of additives sold as such it is intended that regulations implementing EC Directive 89/107/EEC will be made during 1991, replacing the existing provisions of the individual additive regulations.

18. The following gives examples of the labelling details which may appear on some of these foods and which differ from the Food Labelling Regulations requirements:

- *Chocolate Products* – declaration of minimum cocoa solids, trade name and address of registered office.

- *Honey* – name 'honey' is reserved for products meeting a particular standard; indication of type of honey, (comb, chunk etc), controls on further descriptions such as 'clover', 'Mexican', 'West Country' etc.

- *Quality Wines* – vintage, name of vineyard, method of production, region of origin, bottler (there is an exhaustive list of the information which may be given).

- *Hen eggs* – quality grade, size (1, 2, 3 etc) packing date; packing station identity number.

19. The second group of regulations generally introduce labelling requirements additional to the basic provisions contained in the Food Labelling Regulations 1984. These include:

- The Bread and Flour Regulations 1984

- The Butter Regulations 1966 (under review)

- The Caseins and Caseinates Regulations 1985 (implements 83/417/EEC)

- The Cheese Regulations 1970

68

- The Cream Regulations 1970

- The Fruit Juices and Fruit Nectars Regulations 1977 (implements 75/726/EEC)

- The Ice-Cream Regulations 1967 (under review)

- The Margarine Regulations 1967 (under review)

- The Meat Products and Spreadable Fish Products Regulations 1984

- The Natural Mineral Waters Regulations 1985 (implements 80/778/EEC)

- The Salad Cream Regulations 1966 (under review)

- The Skimmed Milk and Non-Milk Fat Regulations 1960

- The Soft Drinks Regulations 1964 (under revision)

20. The following is an example of how the labelling of some of these foods is controlled:

Butter	– reserves name 'butter' to a product meeting a specific compositional standard; requires qualification 'salted' or 'unsalted' as appropriate.
Cheese	– In certain cases, indication of fat content in the name, e.g. 'full fat soft cheese'.
Meat products	– Minimum meat content declaration (immediate proximity to ingredients list), for cured meats and whole meats, maximum added water content declaration in the name of the food.
Natural mineral waters	– Full compositional analysis or compositional statement on compliance with officially recognised analysis; descriptions such as 'naturally carbonated' strictly controlled.

B.   FORTHCOMING AGREED CHANGES

(i)   European Community

Datemarking

1.   Council Directive 89/395/EEC, of 14 June 1989, requires a number of changes to be made to the current datemarking requirements for foods. The Directive proposes a two-tier system of datemarking, with 'best before' for most foods and a compulsory 'use by' date for microbiologically sensitive highly perishable foods. In addition, alternative forms of expressing minimum durability, presently permitted in the Member States (of which the UK 'sell by' date is an example) must be removed by 1 January 1993 at the latest.

2.   The Directive also provides for a uniform Community list of exemptions from the general datemarking requirements. This means that long life and frozen foods, ice cream (apart from individual portions) and cheeses intended to ripen completely or partially in their packaging, currently exempt in the UK, will have to be date marked.

3. Ministers have decided that the changes in paragraph 1 above should be brought in as soon as practicable, in advance of the date for prohibiting trade in non-complying products (20 June 1992). It is proposed that 'use by' dates will therefore be required with effect from *1 April 1991* when 'sell by' dates will also be phased out; 'use by' will be permitted from 1 January 1991. New offences for selling out-of-date 'use by' foods, and for anyone other than the person responsible for the food, to change any datemark are to be introduced. Guidelines have been developed on the type of foods appropriate to the 'use by' date, in consultation with the trade, enforcement interests and Department of Health.

Lot marking

4. Council Directive 89/396/EEC of 14 June 1989, requires all foodstuffs marketed in the Community, with effect from *20 June 1991* to carry an indication of the lot to which a foodstuff belongs (lot/batch mark). The directive leaves it up to the trader to determine the lot and to affix the corresponding mark of his choice, provided this is clearly legible, easily visible and indelible.

Irradiation

5. Council Directive 89/395/EEC, in addition to the above changes, introduces compulsory labelling requirements in respect of irradiated prepacked foods. The requirement is for an indication of treatment to be given in the name of the food using the agreed references 'irradiated' or 'treated with ionising radiation'. Detailed rules for labelling irradiated ingredients of foods will be agreed following the adoption of the EC Directive on Food Irradiation. In the meantime the Food Labelling Directive extends the labelling requirements for foods to food ingredients.

6. Ministers have decided that additional labelling rules in respect of non-prepacked foods and foods sold by caterers should be introduced. The new regulations take effect on 1 January 1991.

Nutrition labelling

7. Council Directive 90/496 provides a two stage approach to nutrition labelling, similar to the Ministry's nutrition labelling guidelines, consisting of a minimum declaration of energy, protein, carbohydrate and fat ('the Big 4'), plus any nutrients, for which a claim is made. After a five year phasing-in period a second list of nutrients to be declared as a group, will be added. This will consist of the 'Big 4' plus sugars, saturates, fibre, and sodium. When a claim or voluntary declaration is made for any of the latter four nutrients, a declaration of all eight nutrients will be required. Starch, sugar alcohols, mono-unsaturates, polyunsaturates, cholesterol and significant amounts of vitamins and minerals can also be declared. Nutrition labelling will be optional except where a nutrition claim is made. A provision has been included in the Directive for an overall review after 8 years.

Food for particular nutritional uses

8. Legislation is needed to implement the PARNUTS Framework Directive 89/398/EEC, in respect of products not to be covered by specific directives and which are not already implemented in UK law. In most respects the Directive repeats provisions contained in Directive 77/94. However, it is necessary to implement the new provisions relating to notification of new products and the power to restrict trade.

Quick frozen foods

9. Council Directive 89/108/EEC, of 21 December 1988, introduces, with effect from 10 January 1991, provisions for the labelling of quick frozen foods (as defined), essentially, if a food is sold under the descriptions 'quick frozen', it must give an indication of minimum durability (datemark) and storage requirements on the label, together with the message 'do not refreeze after defrosting' and a batch number. UK implementing regulations are in preparation.

(ii) **United Kingdom**

Nutrition claims

10. Draft regulations are being drawn up to implement (subject to EC approval) the Food Advisory Committee's recommendations for the control of nutrition claims such as 'low fat', 'sugar free' and 'high fibre'.

Low alcohol

11. Draft provisions to implement the Food Advisory Committee's recommendations to control 'low alcohol' claims have been issued as part of the draft regulations implementing the datemarking and other changes required by Council Directive 89/395. They take effect on 1 January 1991. Changes will also be made to the Food Labelling Regulations as a result of the Licensing (Low Alcohol Drinks) Act 1990 to ensure that the revised licensing laws can be readily enforced and that consumers are not misled.

C. **PROPOSALS UNDER DEVELOPMENT**

(i) **European Community legislation**

1. Proposals for amendment to the EC Food Labelling Directive in relation to three areas are expected shortly:

(a) labelling of single ingredient foodstuffs

(b) labelling of alcoholic drinks (ingredient listing)

(c) percentage indications of characterising ingredients of compound food (QUID)

2. A draft directive on control of *claims* made in the labelling, presentation and advertising of foods is expected shortly. The Commission is committed under the EC Food Labelling Directive, to drawing up a non-exhaustive list of claims, which could mislead the purchaser, and to restricting or prohibiting these as appropriate.

3. Proposals are expected shortly from the Agriculture Directorate (DGVI) relating to *food product quality* issues as a follow up to their 1989 paper on 'The future of rural society'. Indications are that these proposals will include a system on 'appelations d'origine'/geographic indications; an inventory of sales names linked to specifications and a general framework for the use of quality marks on foods.

4. Proposals are currently under development by the Commission relating to the composition and *designation* of *yellow fats*.

5. Nine subordinate directives to be made under the *Foods for Particular Nutritional Uses* (PARNUTS) Directive (89/398/EEC) will lay down specific provisions, which may have labelling implications, for baby, low energy, medical, low-sodium, diabetic, athletes, and gluten free foods.

6. Proposals are under development by the Commission, to control novel foods and novel processes. It is anticipated that these may include labelling provisions.

7. The Commission is committed to proposing controls on 'diet integrators', that is, food supplements such as vitamin and mineral preparations.

(ii) **United Kingdom legislation**

Consideration is being given to the introduction (subject to EC approval) of fish product regulations which would have labelling implications for the names used for these products, and would require fish content declarations.

D. CODES OF PRACTICE

1. A number of codes of practice, usually non-statutory (although they may be used as a point of reference in the Courts), and industry based, are in existence, which may have labelling implications for particular foods. Very often these codes are intended to be an aid to interpretation of legislative provisions, to ensure that a consistent approach is adopted across the sector of industry involved. Alternatively they may serve as a means of regulating and harmonising a particular sector by building on the statutory labelling requirements or to ensure that a particular health/safety message is provided to consumers.

2. Examples are:–

British Meat Manufacturers Association, Code of practice on:
- composition and labelling of meat and meat products
- labelling of reformed cured meat products
- clear and informative labelling of meat products

Dairy Trade Federation, Code of practice on:
- composition and labelling of yogurt

UK Association of Frozen Food Producers, Code of practice on:
- recommended practice for the handling of quick frozen foods (star marking)

Food and Drink Federation, Code of practice on:
- marketing of infant formulae (requires labels to warn against inappropriate preparation of the food)

WHO, Code of practice on:
- marketing of breast milk substitutes (requires labels to warn against inappropriate preparation of the food)

E. GUIDANCE

1. Food Advisory Committee's guidance on the appropriate (and inappropriate) use of 'natural' and related negative claims was accepted by Ministers and issued to all interested parties in June 1989.

APPENDIX IV

THE COMMITTEE'S GUIDELINES ON THE USE OF THE WORD 'NATURAL' AND SIMILAR TERMS

1. The Committee has considered the increasing use of the word 'natural' and words and phrases of like intent in the labelling and advertising of food.

2. The Committee has previously expressed concern that there is a risk of consumers being misled by the increasing use of 'natural' (Final Report on the Review of the Colouring Matter of Food Regulations 1973). It is clear that food manufacturers and traders have seen advantage over the last few years in the wide use of the term, and similar phrases 'no artificial . . .', 'free from artificial . . .' across the whole range of foodstuffs. This is in response to increasing consumer awareness of the diet and its relationship to health, and in particular to the prominence given to concern about the use of additives in food and drink and the high proportion of processed food in the diet. Consumers have been led to accept the idea that food described as 'natural' is somehow of greater worth than food not so described, although there is no inherent reason why this should be so.

3. We had the benefit of a factual survey of the use of 'natural' and similar phrases in the market place commissioned by MAFF from the Local Authorities Co-ordinating Body on Trading Standards, which was considered in detail. We have concluded that the present situation is unsatisfactory. We have confirmed our view that there are many elements of current labelling and advertising which seem to be inaccurate and misleading, and not therefore in the best interest of the consumer or of fair trading. We are also concerned that the concentration on 'naturalness' is diverting attention from more important nutritional messages.

4. The existing framework of law in the Food Act 1984, the Food Labelling Regulations 1984 and the Trade Descriptions Act 1968 provides a broad measure of protection against misleading claims and descriptions. We recognise, however, the problems of enforcement authorities in taking action in an area which involves difficult subjective judgements. The current situation strongly suggests an urgent need to bring about some rationality, but, as our predecessors have recognised*, this is not an area which readily lends itself to the application of detailed regulation. Nevertheless we believe that some guidance on the appropriate use of 'natural' and terms and phrases of similar intent would be welcomed by those responsible for the labelling of food (many of whom have felt the need to follow the trend in order to retain competitiveness), and that enforcement authorities and consumers would also welcome such guidance.

5. Modern agricultural and food manufacturing practices are such a complex combination of nature, craft, art and science that it is not realistic to try to interpret the term 'natural' in a purist sense. Accordingly we have endeavoured to set out some pragmatic advice which is not intended to constitute rigorously defensible definitions of 'natural' or 'artificial' but sets out some broad guidance on the circumstances in which the term 'natural' and similar words and phrases could reasonably be used, and the circumstances in which such terms should not be used.

---

*Food Standards Committee: Report on Claims and Misleading Descriptions 1966
                                     Second Report on Claims and Misleading Descriptions 1980.

This guidance has been expanded and clarified in the light of comments received from the food industry, enforcement authorities and consumers in response to the consultation exercise by the Ministry of Agriculture, Fisheries and Food.

6.   We consider that the principles to apply should be:

(1)   The term 'natural' without qualification should be used only:

(a)   to describe single foods, of a traditional nature to which nothing has been added and which have been subjected only to such processing as to render them suitable for human consumption. We consider freezing, concentration, fermentation, pasteurisation, sterilisation, smoking (without chemicals) and traditional cooking processes such as baking roasting or blanching to be examples of processes which would be acceptable within this criteria. Bleaching, oxidation, smoking (with chemicals) tenderising (with chemicals) and hydrogenation and similar processes would clearly fall outside the criteria. As a general rule we consider that for single ingredient foods, such as cheese, yogurt and butter, acceptable processing would be that which is strictly necessary to produce the final product.

The restriction to food 'of a traditional nature' is intended to exclude foods such as mycoprotein, which may technically be products of natural sources but which we consider do not accord with the public perception of 'natural'.

(b)   to describe food ingredients obtained from recognised food sources and which meet the criteria in (a).

(c)   to describe flavouring substances or permitted food additives obtained from recognised food sources by appropriate physical processing (including distillation and solvent extraction) or traditional food preparation processes. Flavours should only be so described when they are derived wholly from the named food source.

(2)   Compound foods should not therefore be described directly or by implication as 'natural', but it may be acceptable to describe such foods as 'made from natural ingredients' if *all* the ingredients meet the criteria in (1) (b) or (c).

(3)   A food which does not meet the criteria in 6.(1) (a) or (2), should not be claimed to have a 'natural' taste, flavour or colour.

(4)   'Natural', or its derivatives, should not be included in brand or fancy names nor in coined or meaningless phases in such a way as to imply that a food which does not meet the criteria in 6.(1) (a) is natural or made from natural ingredients.

(5)   Claims such as 'natural goodness', 'naturally better', or 'nature's way' are largely meaningless and should not be used.

(6)   'Natural' meaning no more than plain or unflavoured should not be used except where the food in question meets the criteria at 6.(1) (a) or 6.(2).

7.   We are aware that there are other words similar to natural, such as real, genuine, pure, which have separate and distinctive meanings of their own, but which may be used in place of 'natural' in such a way as to imply similar benefits to consumers. In the latter circumstances we consider that the principles set out above should apply equally to these words.

8.   There is in addition a variety of claims (which might be termed 'negative claims') which do not use the term 'natural' or its derivatives, but the effect of which is to imply 'naturalness' to the consumer. We consider such claims potentially misleading and confusing for the consumer and that at least the following should not be used:

    (1)   A claim that a food is 'free from x' if all foods in the same class or category area are free from 'x'.

    (2)   Statements or implication which give undue emphasis to the fact that a product is free from certain non-natural additives, when the product contains other non-natural additives.

    (3)   A claim that a food is 'free from' one category of additive when an additive of another category, or an ingredient, having broadly similar effect is used.

        However we recognise that certain 'negative claims', which do not imply 'naturalness' to the consumer, such as 'free from 'x', where 'x' is a particular additive, may provide accurate and beneficial information for consumers and would not wish to impose unnecessary restrictions in this area.

9.   We believe that the broad guidance set out above might best be taken up voluntarily by those concerned, but if this should not be practicable, consideration should be given to some form of regulation, notwithstanding the difficulties to which we have referred in paragraph 4.

THE COMMITTEE'S REPORT AND RECOMMENDATIONS FOR CONTROLS ON CERTAIN NUTRITION CLAIMS

INTRODUCTION

1.   We considered the need for controls to govern certain popular nutrition claims in food labelling and advertising and the form that such controls might take. There is no intention that our recommendations should set positive standards supplementing the nutrition labelling guidelines, issued by the Ministry of Agriculture, Fisheries and Food. However we recognise that our proposals may have some implications for any system of graphical representations to accompany numerical nutrition declarations; we were aware in making our recommendations that the question of such a system was to be considered separately.

CURRENT LEGISLATION

2.   Misleading claims on food labels are controlled by general provisions in both the Food Act 1984, which requires that food should not be labelled so as to describe it falsely or to mislead the consumer as to its nature, substance or quality, and the Trade Descriptions Act 1968 which prohibits the application of a false trade description. Regulations 36 and 37 and Schedule 6 of the Food Labelling Regulations lay down rules covering specific claims, either express or implied, in the labelling and advertising of food including claims related to the following nutrients: protein, vitamins, minerals, polyunsaturated fatty acids and energy. The basic conditions are that the food must be able to fulfil the claim being made for it, and adequate information has to be given on the label to show the consumer that the claim is justified. Specific conditions relating to the particular claims also apply.

3.   At present there is no specific EC legislation relating to such claims although we understand the Commission intend to make proposals in this area shortly. However, the European Commission has put forward proposals for a Community system of nutrition labelling. These prescribe that any nutrient claim would trigger declarations of energy, protein, carbohydrate, sugars, fat, fibre and sodium and are thus broadly in line with our recommendations.

NEED FOR CONTROLS

4.   The strong consumer interest in diet and health and the response by manufacturers and retailers, many of whom are already providing nutrition information on labels and promoting products on this basis, has resulted in an increase in nutrition claims. These claims have a significant impact on consumers because the message is generally 'flashed' across the front of the food label or package in a prominent position. The main claims of this type which we identified were 'low fat', 'low in saturated fat', 'low salt', 'low sugar', 'no added sugar/salt' 'high fibre' 'more' or 'less' of any of these nutrients and 'free from fat/saturates/sugar/salt'.

5.   We were concerned that this type of claim could mislead consumers because the messages were not consistent across all foods and because the prominence given to such claims could detract from the more important messages about achieving an overall balanced diet. It could also undermine the numerical declarations of nutrients made in accordance with the nutrition labelling guidelines. We believed that consumers had difficulty in assessing the validity of the claims, particularly if they were not backed by further nutrition information on the label, and in comparing both similar foods making similar claims and different foods making similar claims. Breakfast cereals claiming to be 'high-fibre', for example, contain widely different amounts of fibre. Yoghurts and cheese can be found, both bearing the flash 'low fat'; the yoghurt containing about 1% fat and the cheese 15%. We believe that this lack of any standard approach not only creates serious difficulties for the consumer but also causes problems for reputable food manufacturers who want to present their foods as favourably as their competitors are doing but do not want to mislead and run the risk of legal proceedings. In addition we felt that account should be taken of claims which, although factually correct, were misleading because the product contained substances having a similar effect, for example the claim 'no added sugar' when the product contained added glucose, honey or fruit juice.

OPTIONS FOR CONTROLS

6. Having identified a need for controls for those specific nutrition claims which occurred most frequently and were not covered by the Food Labelling Regulations (as listed in paragraph 4) we considered various options. In particular we looked at whether it would be appropriate to make recommendations for guidelines or a code of practice, possibly linking in with the nutrition labelling guidelines, or whether some form of legislation would give better consumer protection. We concluded that legislation was the best means of restricting such claims effectively and that, in view of the precedent of the claims provisions in the Food Labelling Regulations, it would be simplest and clearest to set out these controls in a similar way. We are pleased that this approach has been strongly supported by all sectoral interests, including the trade, although we note that some consumer groups consider that more broadly based action is needed and that this should encompass consumer education. The Committee agrees that there is a need for nutrition education for consumers.

7. The Food Labelling Regulations apply to express and implied claims and it is intended that the additional controls we are proposing will also apply in this way. This intention prompted queries among consultation respondents who were particularly concerned about possible effects on terms such as 'lean meat'. The Committee recognised that it would be difficult to draw a line between what constituted an accurate description of a product and a claim. Much would depend on the context in which the words were used; whether they were intended to influence consumer choice. There would inevitably be borderline cases but the Committee felt that a legitimate distinction could be drawn between acceptable food descriptions (e.g. a butcher stating that a particular cut of meat is 'lean') and a claim which would be covered by our proposals (e.g. Brand x 'LEAN' burgers which strongly suggested a reduced fat claim). We were also aware that no enforcement authority had identified a potential problem with implied claims and we therefore maintain our recommendation that the principle existing in the Food Labelling Regulations be applied.

GENERAL CONDITIONS FOR MAKING CLAIMS

8. We recognised that certain conditions would need to be specific to particular nutrients but that for consistency and clarity a standard approach for each type of claim would be appropriate. The following requirements, which are broadly in line with the existing rules on claims, were considered to be essential general conditions for consistency:

(i) there should be a specified minimum (or maximum) of the relevant nutrient present in the food for absolute claims i.e. 'high'/'low' and relative values for the 'reduced'/'increased' claims;

(ii) consumers should be told about the quantities of nutrients present, including the one(s) for which a claim was made;

(iii) the controls should apply to advertising as well as labelling although detailed nutrition declarations would not have to be made in the advertisement provided the claim itself was substantiated i.e. a 'low fat' claim would trigger fat declarations in the advertisement. It is not intended that the provision should apply to generic advertising (e.g. bread as a source of fibre).

We noted that some 'low x' claims are currently made for products which have a level of 'x' that is quite high, even though it is considerably lower than that of apparently similar products. We believe this leads to serious inconsistencies, which would be avoided if 'low' claims related to a value in grams per 100g of the product. However where products are normally consumed in portions much larger or smaller than 100g, a claim on the 100g basis could be misleading because significantly more or less of the nutrient would be consumed. Whole milk is a good example: on a 100g basis it could claim to be 'low fat', containing 3.9g/100g, but an average serving is twice that volume and 7.8g of fat would contribute a significant quantity of fat to the diet. Therefore, to avoid misleading the consumer, we think *all* products *making a 'low' claim* need to have *no more than the recommended* amount of the relevant nutrient both on a 100g *and* a quantified per serving basis, and in the case of those foods normally consumed in portions significantly larger or smaller than 100g, they should make a declaration relating to both 100g and a specified serving on the label. The serving size would be decided by the manufacturer, but would need to be a reasonable one, or it could be challenged under Section 6 of the Food Act. The Ministry's booklet on 'Food Portion Sizes' gives helpful information on average portion sizes. Both a 'per serving' and a 'reasonable daily intake' basis have precedents in the Food Labelling Regulations and were considered, but we think that the former is likely to be easier to estimate for most foods given the variety of dietary patterns.

9. We had originally proposed that these criteria should apply to 'high' claims also but in the light of responses to the consultation exercise we have concluded that a more flexible approach is needed which would slightly widen the range of foods for which a high fibre claim could be made. This we consider is justified because of the difficulties with the analysis and definition of fibre, the lack of a firm figure for a reasonable daily intake and more importantly because we do not wish to restrict claims to the extent that promotion and consumption of complex carbohydrates is discouraged. We therefore recommend that claims should be based on *either* the amount of fibre in 100g or in a reasonable daily intake whilst retaining the condition that claims should only be permitted for foods which supply a third of the amount suggested, by a number of sources, as a reasonable target for fibre intake per day. If the claim is made on the basis of a reasonable daily intake (or serving size as it will be in many cases) declarations should be made both per 100g and for a specified serving size. For the reasons given in paragraph 8 above where the claim is made for a food normally consumed in much larger or smaller quantities than 100g or 100ml the nutrition declarations under the guidelines should be given on a per serving as well as per 100g basis and the serving size specified.

10. We believe that claims should be allowed for foods for which the absolute 'low' or 'high' claims cannot be made but where an effort has been made to reduce or increase the amount of a nutrient. However such claims should only be in relative terms. To make such a 'reduced' or 'more' claim the food would have to contain a specified percentage of the nutrient greater or less than a standard version of the product. While we recognised that there could be problems in defining what was the standard product, the concept of a 'similar food', typical of those for which no claim is made, as a standard for the dietetic food is already embodied in the diabetic claims section of the Food Labelling Regulations and causes no apparent problems. We felt that a uniform figure of 25% was reasonable for comparative claims. A higher figure would exclude some products which could not technically achieve such a cut without changing their nature but which could contribute significantly to dietary changes. It might also remove the incentive for manufacturers to develop such products. In addition to this condition, in the case of increased fibre claims only, we think it is necessary to specify a minimum fibre content because of the trend of 'fibre' claims on products containing very low amounts of fibre.

11. As indicated in paragraph 8 above we believe that it is necessary when making a claim about a particular nutrient to provide information about the levels of that nutrient present in the food. However we also consider that a consumer seeing a particular nutrition claim might perceive that the food was generally of nutritional benefit and might not, for example, expect a 'low fat' product to be high in salt. Whilst we do not think that it would be reasonable to draw up conditions on this basis we believe that the giving of adequate nutrition information should be obligatory when a nutrition claim is made so that the consumer has the full nutritional picture for that food. As all the nutrients we have covered are ones which figure in the guidelines on nutrition labelling, issued by the Ministry of Agriculture, Fisheries and Food, we recommend that a claim should trigger a Category III declaration in accordance with these guidelines. Levels of all the following nutrients would therefore have to be given: energy, protein, carbohydrate (with a breakdown to show sugars), fat (with a breakdown to show saturates), sodium and fibre. As stated in paragraph 9 above, we believe that it is helpful for the consumer for the declarations to be given on a per serving as well as per 100g where claims, e.g. for fibre, are made on the basis of a serving or reasonable daily intake. The serving size should be specified. We recommend that these conditions should also apply to existing provisions controlling nutrition claims.

12. We consider that foods naturally low or high in nutrients should be permitted to make nutrition claims provided they satisfy the relevant requirements. but must do so in a way which makes it clear that all foods of that kind are similar, and there is nothing about the nutrient level which is specific to a particular brand. There is a precedent for this in the 'low energy' claims section of the Food Labelling Regulations and we therefore recommend the use of the term 'a . . . food'. For example frozen peas could claim to be 'a low salt food'.

13. We also looked at the possibility of prohibiting claims being made for foods the 'normal' serving size of which would contribute a negligible quantity of nutrients to the diet, for example certain sauces and condiments. However it was felt that manufacturers would benefit little by making nutrient claims. particularly of the 'low', 'high'. 'reduced', 'increased' variety, which made most impact, for products normally consumed in very small quantities and whose contribution to the diet was therefore insignificant. There are relatively few such claims and so long as they can be justified we believe that they are unlikely to mislead the consumer.

SPECIFIC CONDITIONS – LEVELS LIMITING CLAIMS

14. As a general principle we believe that the levels chosen should be such as to enable consumers to make significant increases or decreases in their consumption of particular nutrients. Lax rules, particularly for the absolute claims, would bring the concept into disrepute. However we also took account of the practical difficulties involved in cutting back on the amount of certain ingredients containing a particular nutrient. We did not wish to propose rules so strict that manufacturers would be discouraged from making any adjustments at all to their products because they could not hope to reach an unrealistically low level where a claim could be made. With such a diverse array of nutrients, we considered that it was not possible to choose one overriding principle justifying selection of the absolute levels for 'low'. 'high' claims and the choice is essentially pragmatic. We looked at the possibility of using a percentage energy basis, related to the recommended levels of intake arising from the advice of the Committee on Medical Aspects of Food Policy and from other reports, but decided that whilst such a concept was relevant to fat, for example, where recommendations were in terms of percentage energy contribution, it was not appropriate for sodium, or sugars – where cariogenic effect is as likely to be related to the absolute amount taken, and to patterns of intake, as to the proportion of it in the diet.

15. In setting the levels for 'high'/'low' claims we were therefore aiming for a largely standard approach, taking account of consumer perceptions, practical feasibility for manufacturers and nutrient levels in foods currently making claims. For fat and sugar we recommend that to make a 'low' claim the nutrient content of 100g of the food and in a normal serving size should not exceed 5g and for saturates the limit should be 3g per 100g. The minimum quantities recommended to control 'high fibre' (6g per 100g), 'source of fibre' (3g per 100g) claims were related to the figure of 18g suggested by a number of sources as a reasonable target for fibre intake per day. The figure of 40mg per 100g as the level for a 'low salt' claim was based on an American limit. We recognise that this is stringent but we considered that such a limit would be helpful for those suffering from high blood pressure who are on low sodium diets. The limits proposed for 'free from' claims are deliberately restrictive because such a claim should normally mean that the nutrient is not detectable in the food. However we recognised that it would be difficult to exclude the nutrient completely and therefore, for practical reasons, we suggest an upper limit of 0.2g/100g for sugar, 0.1g/100g for saturates, 0.15g for fat and 5mg/100g for sodium.

16. We recognise that the limit for a 'low in saturates' claim is out of line with the current provisions in the Food Labelling Regulations, for polyunsaturated fatty acids claims, aimed essentially at oils and fats. These require that the claim must be accompanied by the words 'low in saturates' but the condition is that not more than 25% of the fatty acids may be saturated. We believe that the 25% saturates condition remains valid for oils and fats for which polyunsaturates claims are made but is not appropriate for saturated fatty acids claims covering the whole range of foods. We recommend deletion of the requirement to include a 'low in saturates' statement with a polyunsaturates claim.

SUMMARY OF RECOMMENDATIONS

17. Our detailed recommendations are set out in the attached Annex which has been drawn up along the lines of Schedule 6 of the Food Labelling Regulations 1984. The following is a summary of our advice:

(a) New legislative controls should be introduced for certain nutrition claims following the precedents set by existing provisions in the Food Labelling Regulations 1984. They should thus apply to express and implied claims and to claims made in advertising as well as in food labelling (paragraphs 4 – 7).

(b) The conditions for making claims should apply to advertisements but the declarations required need only to be sufficient to substantiate the claim. However, detailed nutrition declarations would have to be given on the label when the food is sold (paragraph 8).

(c) The range of claims to be covered should be 'low fat' 'low in saturated fat' 'low salt' 'low sugar' 'no added sugar/salt' 'high fibre', 'more' or 'less' of any of these nutrients and 'free from fat/saturates/sugar/salt' (paragraph 4).

(d) The restriction on 'low' claims should be set in terms of absolute limits for each nutrient, and products making such claims should have to meet the limit an both a grams per 100 grams and on a per normal serving size basis (paragraph 8).

(c) The restriction on 'high fibre' claims should be set in terms of an absolute limit and products making such claims should meet the limit on either a grams per 100g or on a reasonable daily intake basis (paragraph 9).

(f) For comparative claims a 25% reduction/increase should be required by comparison with the 'normal' product (paragraph 10).

(g) In the case of 'no added' claims for sugar and salt the food should not be permitted to contain other similar substances (paragraph 5).

(h) A nutrition claim should trigger full nutritional declarations in accordance with the guidelines on nutrition labelling issued by the Ministry of Agriculture Fisheries and Food and where the claim is based on a serving or reasonable daily intake of the food, as it may be for certain fibre claims, the nutrition declarations should be given for a specified serving size as well as on a grams per 100g basis. These conditions should be extended to existing nutrition claims provisions in the Food Labelling Regulations (paragraphs 8, 9 and 11).

(i) When an absolute claim is made for a food normally consumed in significantly larger or smaller quantities than 100g or 100ml the nutrition declarations should be given for a specified serving size as well as on a grams per 100g basis (paragraphs 8 and 9).

(j) If the claim relates to a food which is naturally high or low in nutrients this should be made clear (paragraph 12).

(k) The requirement in the Food Labelling Regulations that a 'low in saturates' statement should be given when a claim for polyunsaturates is made should be deleted (paragraph 16).

ANNEX

RECOMMENDED CONDITIONS FOR MAKING PARTICULAR NUTRITION CLAIMS

(The references to nutrition declarations mean a declaration on the basis of the guidelines on nutrition labelling issued by the Ministry of Agriculture, Fisheries and Food).

(1) A claim of a type described in Column 1 shall not be made, either expressly or by implication, in the labelling or advertising of a food except in accordance with the conditions specified in Column 2.

(2) Where a claim is a claim of two or more of the types described in Column 1, the conditions appropriate to each of the relevant types of claim shall be observed.

Claim	Conditions
1. Total fat A claim that a food has a *reduced fat content*	1. Total fat content of 100g (or 100ml) of the food must not be more than three quarters of that of 100g (or 100ml) of a similar food typical of those for which no such claim is made. 2. A Category III nutrition declaration must be made. 3. Where the claim is made in the advertising of a food the claim must be substantiated in the advertisement and the food when sold must be labelled with the particulars specified above.
A claim that a food is *low in fat*	1. (a) Total fat content of 100g (or 100ml) of the food must not exceed 5g *and* (b) total fat content of a normal serving of the food must not exceed 5g 2. Where a food is naturally low in fat in the above terms, the claim must be made in the form 'A low fat food'. 3. (a) A Category III nutrition declaration must be made. (b) Where the claim is made in respect of a food, the normal serving of which is less than 50g/ml or more than 150g/ml, the declaration specified in (a) must be made in addition for a given serving of the food. The serving size must be specified. 4. Where the claim is made in the advertising of a food the claim must be substantiated in the advertisement and the food when sold must be labelled with the particulars specified above.

Claim	Conditions
A claim that a food is *fat free*	1. Total fat content as appropriate of 100g (or 100ml) of the food must not exceed 0.15g. 2. A Category III nutrition declaration must be made. 3. Where the claim is made in the advertising of a food the claim must be substantiated in the advertisement and the food when sold must be labelled with the particulars specified above.
2. Saturates A claim that a food has *reduced content of saturates**	1. Total saturates content of 100g (or 100ml of the food must not be more than three quarters of that of 100g (or 100ml) of a similar food typical of those for which no such claim is made. 2. A Category III nutrition declaration must be made. 3. Where the claim is made in the advertising of a food the claim must be substantiated in the advertisement and the food when sold must be labelled with the particulars specified above.
A claim that a food is *low in saturates**	1. (a) Total saturates content of 100g (or 100ml) of the food must not exceed 3g *and* (b) total saturates content of a normal serving of the food must not exceed 3g. 2. Where a food is naturally low in saturates in the above terms, the claim must be made in the form 'A low saturates food'. 3. (a) A Category III nutrition declaration must be made. (b) Where the claim is made in respect of a food, the normal serving of which is less than 50g/ml or more than 150g/ml, the declaration specified in (a) must be made in addition for a given serving of the food. The serving size must be specified. 4. Where the claim is made in the advertising of a food the claim must be substantiated in the advertisement and the food when sold must be labelled with the particulars specified above.

* As defined in the Ministry's revised nutrition labelling guidelines.

Claim	Conditions
A claim that a food is *saturates free**	1. Total saturates content of 100g (or 100ml) of the food must not exceed 0.1g.
	2. A Category III nutrition declaration must be made.
	3. Where the claim is made in the advertising of a food the claim must be substantiated in the advertisement and the food when sold must be labelled with the particulars specified above.

3. Sugars

Claim	Conditions
A claim that a food has a *reduced sugar/s content.* Note: The use of the term 'reduced' when applied to a jam or similar product in accordance with the Jams and Similar Products Regulations 1981 (as amended) shall not itself constitute a claim of the type described.	1. Total sugars content of 100g (or 100ml) of the food must not be more than three quarters of that of 100g (or 100ml) of a similar food typical of those for which no such claim is made.
	2. A Category III nutrition declaration must be made.
	3. Where the claim is made in the advertising of a food the claim must be substantiated in the advertisement and the food when sold must be labelled with the particulars specified above.

Claim	Conditions
A claim that a food is *low in sugar/sugars*	1. (a) Total sugars content of the food must not exceed 5g per 100g (or 100ml).
	and
	(b) the total sugars value of a normal serving of the food must not exceed 5g.
	2. In the case of a food which is naturally low in sugars in the above terms, the claim must be in the form of 'A low sugar food'.
	(a) A Category III nutrition declaration must be made.
	(b) Where the claim is made in respect of a food, the normal serving of which is less than 50g/ml or more than 150g/ml, the declaration specified in (a) must be made in addition for a given serving of the food. The serving size must be specified.
	4. Where the claim is made in the advertising of a food the claim must be substantiated in the advertisement and the food when sold must be labelled with the particulars specified above.

* As defined in the Ministry's revised nutrition labelling guidelines.

Claim	Conditions
A claim that a food is *sugar/s free*	1. Total sugars content of 100g (or 100ml) of the food must not exceed 0.2g. 2. A Category III nutrition declaration must be made. 3. Where the claim is made in the advertising of a food the claim must be substantiated in the advertisement and the food when sold must be labelled with the particulars specified above.
A claim that a food has *no added sugar/sugars or is unsweetened*. Note: The use of the term 'unsweetened' in accordance with the provisions of the Condensed Milk and Dried Milk Regulations 1977 (as amended) shall not of itself constitute a claim of the type described.	1. No mono- or disaccharides or food composed mainly of these sugars should have been added to the food or to any of its ingredients. 2. A Category III nutrition declaration must be made. 3. Where the claim is made in the advertising of a food the claim must be substantiated in the advertisement and the food when sold must be labelled with the particulars specified above.
4. Fibre A claim that a food contains *more/increased/higher fibre**	1. Total fibre content of 100g (or 100ml) of the food must be at least 25% more than that of 100g (or 100ml) of a similar food typical of those for which no such claim is made *and* 2. total fibre content of 100g (or 100ml) or of a reasonable daily intake of the food for which the claim is made must be at least 3g. 3. (a) A Category III nutrition declaration must be made. (b) Where the claim is made in respect of a food, the normal serving of which is less than 50g/ml, or more than 150g/ml, the declaration specified in (a) must be made in addition for a given serving of the food. The serving size must be specified. 4. Where the claim is made in the advertising of a food the claim must be substantiated in the advertisement and the food when sold must be labelled with the particulars specified above.

* Calculated as non-starch polysaccharides.

Claim	Conditions
A claim that a food is a *source of fibre**	1. (a) Total fibre content of 100g (or 100ml) of the food must be at least 3g *or* (b) total fibre content of a reasonable daily intake of the food should be at least 3g. 2. A Category III nutrition declaration must be made for 100g (or 100ml) and – where the claim is made on the basis of a reasonable daily intake or serving – the declaration must be made in addition for a given serving of the food. The serving size must be specified. 3. Where the claim is made in the advertising of a food the claim must be substantiated in the advertisement and the food when sold must be labelled with the particulars specified above.
A claim that a food is *high/rich in fibre**	1. (a) Total fibre content of 100g (or 100ml) of the food must be at least 6g or (b) total fibre content of a reasonable daily intake of the food must be at least 6g. 2. In the case of a food naturally high in fibre in the above terms, the declaration should take the form 'A high fibre food'. 3. (a) A Category III nutrition declaration must be made for 100g (or 100ml) and – where the claim is made on the basis of a reasonable daily intake or serving – the declaration must be made in addition for a given serving of the food. The serving size must be specified. (b) Where the claim is made in respect of a food, the normal serving of which is less than 50g/ml or more than 150g/ml, the declaration must be made for a given serving of the food in addition to the declaration on 100g/ml basis. The serving size must be specified. 4. Where the claim is made in the advertising of a food the claim must be substantiated in the advertisement and the food when sold must be labelled with the particulars specified above.

* Calculated as non-starch polysaccharides.

Claim	Conditions
5. Salt A claim that a food has a *reduced salt/sodium content*	1. Total sodium content of 100g (or 100ml) of the food must not be more than three quarters of that of 100g (or 100ml) of a similar food typical of those for which no such claim is made. 2. A Category III nutrition declaration must be made. 3. Where the claim is made in the advertising of a food the claim must be substantiated in the advertisement and the food when sold must be labelled with the particulars specified above.
A claim that a food is *low in salt/sodium*	1. (a) The total sodium content of 100g (or 100ml) of the food must not exceed 40mg *and* (b) the total sodium content of a normal serving of the food must not exceed 40mg. 2. In the case of a food which is naturally low in sodium in the terms of these regulations, the claim must be in the form of 'A low salt/sodium food'. 3. (a) A Category III nutrition declaration must be made. (b) Where the claim is made in respect of a food, the normal serving of which is less than 50g/ml or more than 150g/ml, the declaration specified in (a) must be made in addition for a given serving of the food. The serving size must be specified. 4. Where the claim is made in the advertising of a food the claim must be substantiated in the advertisement and the food when sold must be labelled with the particulars specified above.
A claim that a food is *sodium/salt free*	1. Total sodium content of 100g (or 100ml) of the food must not exceed 5mg. 2. A Category III nutrition declaration must be made. 3. Where the claim is made in the advertising of a food the claim must be substantiated in the advertisement and the food when sold must be labelled with the particulars specified above.

Claim	Conditions
A claim that a food contains *no added salt/sodium*	1. No salt or other salts of sodium should have been added to the food or to any of its ingredients. 2. A Category III nutrition declaration must be made. 3. Where the claim is made in the advertising of a food the claim must be substantiated in the advertisement and the food when sold must be labelled with the particulars specified above.

Footnote: Where a food is in a concentrated or dehydrated form and is intended to be reconstituted by the addition of water or other substances, the conditions set out in Column 2 above shall apply to the food when reconstituted as directed.

APPENDIX VI

REVISED OCTOBER 1990

THE COMMITTEE'S GUIDELINES FOR THE LABELLING OF FOODS PRODUCED USING GENETIC MODIFICATION

The Food Advisory Committee has been considering the role which genetic engineering is beginning to play in food production and has been concerned to identify those uses of the technology which could present moral and ethical concerns for some consumers. The Committee has concluded that there are some uses of the technology in the production of food which would generate such concerns because of the possible presence of genetically modified organisms (GMOs) in the final food. The Committee felt that labelling was not the answer to these concerns except where the presence or use of GMOs could be considered to alter materially the nature of the food. The Committee concluded that under these circumstances specific food labelling should be used to inform consumers of the use of gene technology. The Committee has designated four basic food categories as a primary screening mechanism to determine when specific food labelling is required. These four food categories which have formed the basis of the Guidelines for the Labelling of Foods Produced Using Genetic Modification, have been developed to assist the Committee with its own work. Therefore it should not be assumed that the labelling advice for each of the four categories would automatically apply in every case. The Committee wishes to consider the labelling requirements for such foods on a case-by-case basis and as a consequence these guidelines may then need to be revised. The four categories of the primary screen are defined below:

1. *Nature identical food products of GMOs*

   Foods which are the products of, or which contain products of, a genetically modified organism (but not the organism itself, its cells or DNA) which are identical to products from conventional organisms traditionally consumed in Western Europe.

   This category refers only to the use of products of a genetically modified organism in food and not the organism itself, its cells or its DNA, and furthermore restricts itself to products which are identical to those which are traditionally consumed in Western Europe and are derived by conventional methods from conventional materials. In this case the Committee already has a precedent. When this Committee considered the case of need for the enzyme chymosin, produced from the transfer of the calf chymosin gene to a micro-organism, it was concluded that because the product, bacterial chymosin B, was indistinguishable from calf chymosin B, it would be unnecessary to require special labelling provisions. The Committee confirmed this view as a general principle for this category of products.

2. *Food from intra-species GMOs*

   Foods which are or which contain a genetically modified organism (or its cells and/or DNA) which is produced only from the gene pool of its own species.
   In this category the process of genetic modification can be said to represent a more rapid and effective alternative to selective breeding programmes for

89

achieving the required characteristics of the organism. The Committee felt that in the majority of cases in this category, special labelling would be unnecessary, but each case should be considered on its merits. The essential point for consideration was whether the consumer was likely to be misled, or might wish to be informed. The Committee when subsequently asked to consider special labelling provisions for a genetically modified bakers yeast, which resulted from the insertion of a gene from another variety of yeast, decided that this organism clearly fell into this category and that it was not necessary to apply special labelling provisions to the bread which contains it.

3. *Novel food products of GMOs*

Foods which are the products of, or which contain products of a genetically modified organism (but not the organism itself, its cells or DNA) which differ from products from conventional organisms traditionally consumed in Western Europe.

The products in this category are not identical to products currently in the diet from conventional organisms. They may include for example novel fats, proteins or carbohydrates produced by GMOs. The Committee stated that as a general principle special labelling of food which contains these products of GMOs would probably be required.

4. *Foods from trans-species GMOs*

Foods which are or which contain a genetically modified organism (or its cells and/or DNA) which is produced by the introduction of genetic material from any source other than the same species as the host organism excluding non functional short chain DNA linker sequences.

The Committee generally felt that the greatest potential source of public concern about genetically modified organisms lies with this category, since the incorporation of genes from one species into a host organism of a different species represents a departure from that which can be achieved by conventional breeding practices. It was recognised that the end-product, a trans-species genetically modified organism, would provide a focus of moral and ethical concern for some consumers and that sufficient information should be provided through food labelling in order that consumers can choose to avoid such products if they wish. The Committee's recommendations on the labelling of foods, which have been subjected to irradiation were cited as an example of such an approach. For cases falling into this category, the Committee therefore agreed to the principle that there would probably be a need for labelling in *all* cases to allow consumers to distinguish foods made from these organisms from foods made with conventional organisms, however, the Committee will consider each case on its merits. The Committee accepted that although foods from trans-species GMOs would normally require specific labelling, where such GMOs derive from closely related species which already produce fertile off-spring by natural means, e.g a GMO loganberry from blackberry and raspberry parents, an exception might be made, but that such decisions would be made on a case-by-case basis.

The form of the labelling declaration

When consumers should be informed about the use of GMOs and their products in, or as foods the Committee has decided that the following statement should be made as part of the labelling requirements:

'(contains) products of gene technology'

and that when an ingredient of the food is a product of gene technology, this should also be identified in the ingredients list.

1. For the purposes of this document the term genetically modified organism is defined as:

'organisms in which the genetic material has been altered in a way that does not occur naturally by mating and/or natural recombination'.

Included within the definition are the following techniques:

(a) recombinant DNA techniques using vector systems as previously covered by Council Recommendation 82/472/EEC. [ie the formation of new combinations of genetic material by the insertion of nucleic acid molecules produced by whatever means outside the cell, into any virus, bacterial plasmid or other vector system so as to allow their incorporation into a host organism in which they do not naturally occur but in which they are capable of continued propagation].

(b) techniques involving the direct introduction into a (micro-) organism of heritable material prepared outside the (micro-) organism including micro-injection, macro-injection and micro-encapsulation.

(c) cell fusion or hybridisation techniques where live cells with new combinations of heritable genetic material are formed through the fusion of two or more cells by means of methods that do not occur naturally.

Techniques which are not considered to result in genetic modification on the condition that they do not involve the use of r-DNA molecules or GMOs are:

(a) in vitro fertilisation

(b) conjugation, transduction, transformation or any other natural process

(c) polyploidy induction

For the purposes of this document the following techniques are not considered to result in genetic modification, provided that they do not involve the use of GMOs as recipient or parental organisms:

(a) mutagenesis

(b) cell fusion (including protoplast fusion) of cells from plants which can be produced by traditional breeding methods.

APPENDIX VII

GLOSSARY OF TERMS AND ABBREVIATIONS

ACP   – Advisory Committee on Pesticides

ADI   – Acceptable Daily Intake

CMO   – Chief Medical Officer

COI   – Central Office of Information

COMA   – Committee on Medical Aspects of Food Policy

FAWC   – Farm Animal Welfare Council

FDA   – US Food and Drugs Administration

GMO   – Genetically Modified Organism

HEA   – Health Education Authority

MRL   – Maximum Residue Limit

PARNUTS – Foods for Particular Nutritional Uses

PAS   – Public Attitude Surveys Limited

QUID   – Quantitative Declaration of Ingredients

Printed in the United Kingdom for HMSO Dd.0293749 4/91 C20 488/2 12521